Lecture Notes in Computer Science 5019

Commenced Publication in 1973
Founding and Former Series Editors:
Gerhard Goos, Juris Hartmanis, and Jan van Leeuwen

Jose A. Onieva   Damien Sauveron
Serge Chaumette   Dieter Gollmann
Konstantinos Markantonakis (Eds.)

Information Security Theory and Practices

Smart Devices, Convergence and Next Generation Networks

Second IFIP WG 11.2 International Workshop, WISTP 2008
Seville, Spain, May 13-16, 2008
Proceedings

 Springer

Volume Editors

Jose A. Onieva
University of Malaga, 29071 Malaga, Spain
E-mail: onieva@lcc.uma.es

Damien Sauveron
Université de Limoges/CNRS, 87000 Limoges, France
E-mail: damien.sauveron@xlim.fr

Serge Chaumette
Université Bordeaux 1/CNRS, 33405 Talence Cedex, France
E-mail: Serge.Chaumette@labri.fr

Dieter Gollmann
Hamburg University of Technology, 21071 Hamburg, Germany
E-mail: diego@tu-harburg.de

Konstantinos Markantonakis
University of London, Egham,TW20 0EX, UK
E-mail: k.markantonakis@rhul.ac.uk

Library of Congress Control Number: 2008926221

CR Subject Classification (1998): E.3, K.6.5, C.3, D.4.6, C.2

LNCS Sublibrary: SL 4 – Security and Cryptology

ISSN   0302-9743
ISBN-10   3-540-79965-6 Springer Berlin Heidelberg New York
ISBN-13   978-3-540-79965-8 Springer Berlin Heidelberg New York

Springer is a part of Springer Science+Business Media

springer.com

© IFIP International Federation for Information Processing, Hofstrasse 3, A-2361 Laxenburg, Austria 2008

Typesetting: Camera-ready by author, data conversion by Scientific Publishing Services, Chennai, India
Printed on acid-free paper   SPIN: 12269297   06/3180   5 4 3 2 1 0

Preface

With the rapid development of information technologies and the transition to next-generation networks, computer systems and in particular embedded systems are becoming more and more mobile and ubiquitous. They also strongly interact with the physical world. Ensuring the security of these complex and resource-constrained systems is a really challenging research topic. Therefore this Workshop in Information Security Theory and Practices was organized to bring together researchers and practitioners in related areas, and to encourage cooperation between the research and the industrial communities.

This was the second edition of WISTP, after the first event in Heraklion, Greece, in 2007. This year again we had a significant number of high-quality submissions coming from many different countries. These submissions reflected the major topics of the conference, i.e., smart devices, convergence, and next-generation networks. Submissions were reviewed by at least three reviewers, in most cases by four, and at least by five for the papers involving Program Committee members. This long and rigorous process could be achieved thanks to the hard work of the Program Committee members and additional reviewers, listed in the following pages. This led to the selection of high-quality papers that made up the workshop program and are published in these proceedings. A number of posters and short papers were also selected for presentation at the conference.

The process was very selective and we would like to thank all those authors who submitted contributions that could not be selected. We also want to acknowledge the great involvement of the keynote speakers who contributed to making WISTP a valuable event.

As a collocated event, we were happy to host a Segur@ project meeting. The Segur@ project is an initiative of Telefonica, which is the consortium leader, together with a total of 12 of the major Spanish information systems security and telecommunications companies. The main goal of the project is to generate a trust and security framework to foster the use of information technologies and communications (ITC) in the e-society. A number of public research bodies, subcontracted by the participating partners, also take part in the project.

To host a successful workshop requires not only support from the research community but also practical and financial support from a range of companies and scientific organizations, which we would like to thank. We have to thank all those who have been involved in the different committees and at different points in the organization process.

We sincerely hope that all attendees enjoyed the scientific contents of the workshop and the networking opportunity which is one of the strong characteristics offered by WISTP.

March 2008

Serge Chaumette
Dieter Gollmann

Organization

WISTP 2008 was organized by the University of Malaga and XLIM (University of Limoges, CNRS).

Steering Committee

Angelos Bilas, FORTH-ICS and University of Crete, Greece
Konstantinos Markantonakis, ISG-SCC, Royal Holloway University of London, UK
Jean-Jacques Quisquater, DICE, Catholic University of Louvain, Belgium
Damien Sauveron, XLIM, (University of Limoges, CNRS), France

General Chairs

Jose A. Onieva, Computer Science Department, University of Malaga, Spain
Damien Sauveron, XLIM, (University of Limoges, CNRS), France

Workshop/Panel/Tutorial Chair

Konstantinos Markantonakis, ISG-SCC, Royal Holloway
University of London, UK

Local Organizers

Diego R. Lopez, Red.es - RedIRIS, Spain
Javier Lopez, Computer Science Department, University of Malaga, Spain

Publicity Chairs

Claudio Ardagna, Department of Information Technologies,
University of Milan, Italy
Samia Bouzefrane, CEDRIC, CNAM, France
Joonsang Baek, Institute for Infocomm Research (I2R), Singapore

Program Committee

Chairs: Serge Chaumette, LaBRI, University Bordeaux 1, France
Dieter Gollmann, Security in Distributed Applications,
TU Hamburg-Harburg, Germany

Damien Sauveron, XLIM, (University of Limoges, CNRS), France
Willy Susilo, CCISR, University of Wollongong, Australia
Michael Tunstall, University College Cork, Ireland
Paulo Jorge Esteves Veríssimo, LASIGE, University of Lisbon, Portugal

Additional Referees

Jérémie Albert
Eve Atallah
Mika Cohen
Ton van Deursen
Gerhard Hancke
Martin Johns
Giorgos Karopoulos
Achraf Karray
Olivier Ly
Ilaria Matteucci

Jun Pang
Fabio Picconi
Nicolas Prigent
Sasa Radomirovic
Peter van Rossum
Georgios Spathoulas
Alexandros Tsakountakis
Claus Wonnemann
Jan Zibuschka

Sponsoring Institutions

University of Malaga
XLIM (UMR Université de Limoges/CNRS 6172).

Main Sponsors

Since the early stages of the workshop inception the workshop organizers received positive feedback from a number of high-profile organizations, that further led to direct financial support. This enabled the workshop organizers to strengthen significantly their main objective for proposing a high-standard academic workshop. The support helped significantly to keep the workshop registration costs as low as possible and at the same time offer a number of best paper awards. Therefore, we would like to express our gratitude and thank every single organization. We are also looking forward to working together in future WISTP events.

Table of Contents

Untraceability of RFID Protocols

Ton van Deursen, Sjouke Mauw, and Saša Radomirović

Université du Luxembourg
Faculté des Sciences, de la Technologie et de la Communication
6, rue Richard Coudenhove-Kalergi
L-1359, Luxembourg

Abstract. We give an intuitive formal definition of untraceability in the standard Dolev-Yao intruder model, inspired by existing definitions of anonymity. We show how to verify whether communication protocols satisfy the untraceability property and apply our methods to known RFID protocols. We show a previously unknown attack on a published RFID protocol and use our framework to prove that the protocol is not untraceable.

Keywords: RFID protocols, untraceability, formal verification.

1 Introduction

Radio frequency identification (RFID) systems aim to identify tags to readers in an open environment. Communication between readers and tags is even possible when there is no physical or visual contact between tags and readers. RFID tags can be very small and cheap [1] and can therefore be embedded in a wide variety of objects. They have, for instance, been embedded in passports [2] and there are plans to embed them in bank notes [3] and groceries [4,5].

The absence of physical contact during communication and the expected ubiquity of RFID systems will only encourage nefarious entities to trace and observe tags through time and space. If at any such point a tag is linked to a person, the tracing of a tag becomes the tracing of a person.

The need for RFID protocols to be resistant against this kind of attack on privacy has been realized early on. Intuitively, protocols are said to provide untraceability, if an adversary is not able to recognize a tag he previously observed. Although untraceability is mainly mentioned in the context of RFID systems, it is an issue for any protocol which is used with a mobile device. In the Bluetooth setting, it is known as location privacy [6,7].

History has shown that designing protocols is a difficult and error-prone task and that formal verification of security properties is necessary [8,9]. While traditional security properties such as authentication and secrecy have been studied thoroughly, untraceability has only become relevant with the introduction of travelling devices. Until now it has typically been treated rather informally. In some cases, protocol designers prove untraceability of their protocols without even defining it properly.

J.A. Onieva et al. (Eds.): WISTP 2008, LNCS 5019, pp. 1–15, 2008.

In this paper, we propose an intuitive, formal definition for untraceability that is inspired by existing definitions for anonymity [10,11]. We demonstrate the usability of our definition on two protocols. In particular, we prove that the mutual authentication protocol by Feldhofer, Dominikus, and Wolkerstorfer [12] is untraceable and that the Di Pietro and Molva protocol [13] is untraceable only for a *restricted choice of parameters* and assuming that their constructed function, DPM, is a *perfect hash function*. By removing the assumptions and analyzing the algebraic properties of the DPM function we demonstrate the first, efficient method to trace tags running the Di Pietro-Molva protocol. We then relate this insight back to our definition of untraceability by exhibiting a trace of the protocol which violates our definition.

Our paper is structured as follows. In the next section we discuss related work. In Section 3 we formally define untraceability. In Section 4 we prove the Feldhofer *et. al.* protocol untraceable and in Section 5 we discuss the Di Pietro-Molva protocol. We conclude with an outlook on future work in Section 6.

2   Related Work

A discussion of the importance of untraceability can be found in [14,15,16,17,18]. Several RFID protocols have been proposed with informal reasoning about their untraceability property [19,20,21,22] or based on the belief that protocols with random nonces in all messages are untraceable [23,24,25]. On the opposite end of the spectrum, pseudonyms and frequent changes of IDs are claimed to be necessary to avoid the tracking problem [26,27,28,29]. Among the cryptographic notions of untraceability, worth mentioning are [30,31,32,33,34,35,36].

The notion of untraceability defined in this paper is stronger than the notions considered in [37,38] in the following sense. These works consider RFID tags which an adversary could recognize between any two successful communications with a trusted RFID reader to be untraceable, while under the present definition they are not.

The untraceability notion considered here is only weakly related to the unlinkability notion that has been studied extensively in privacy enhancing technologies (PET) literature. A formal definition for unlinkability was given in [39,40]. Unlinkability considers whether links can be established between senders and receivers, while untraceability considers whether different communications can be attributed to the same agent. It is difficult to give a precise relation between anonymity and untraceability due to the many differing definitions of anonymity [10,11,41]. In general however, untraceability implies anonymity.

Security properties such as secrecy and authentication, implemented by a protocol at a certain layer, are maintained in the lower layers. However, for untraceability, the property can be compromised by the protocols in the lower layer [42]. In this paper, we will focus on untraceability in the application layer.

3   Definitions

3.1   Security Protocol Model

The purpose of this section is to introduce basic notation and definitions concerning security protocols. Rather than providing a full description of security protocol syntax and semantics, we will only present the basic requirements needed for defining and analyzing untraceability. In short, we require that the behavior of a number of agents executing a security protocol is described by a set of traces in which we can identify the events belonging to the same run. A full semantics satisfying our requirements can be found in [43].

The starting point is the specification of a security protocol. A security protocol defines the behavior of a set of *roles* (e.g. initiator, responder, server). A role specification consists of a sequence of events (e.g. the sending or reception of a message). The messages contained in the events are *role terms*. Role terms are built from *basic role terms*, such as nonces (typically denoted by n), role names (e.g. R or T), or keys (typically denoted by k). Complex terms can be constructed using functions, such as tupling (denoted by $(t_1, \ldots, t_n)$), encryption ($\{t\}_k$), hashing ($h(t)$), and exclusive or (denoted by $\oplus$). Throughout this paper we will use Message Sequence Charts to present security protocol specifications (see e.g. Figure 1 in Section 4). In such a diagram, we use a hexagon at the end of a role specification to denote a security claim, such as untraceability.

A role specification is only a blueprint of some actual behavior. It serves as the program that an agent (typically denoted by *Alice* or *Bob*) can execute. An execution of a role R specified by a protocol P is called a *run* of R. Such a run will be denoted by $R\#\theta$, where θ is a (unique) run identifier. A run can thus be viewed as an instantiation of a role. Therefore, we will also have to instantiate the abstract role events, yielding the *run events*. Run events are constructed from role events by instantiating the contained role terms. An instantiated role term is a *run term*. Run terms are similar to role terms, except that roles are replaced by agents and that basic role terms are suffixed with the identifier of the run. An instantiated nonce n is denoted by $n\#\theta$ if it occurs in run $R\#\theta$. In this way occurrences of the same nonce in different runs can be distinguished.

We assume a standard Dolev-Yao adversary, characterized by its *knowledge*. This knowledge consists of the set of run terms that the adversary initially knows, extended with the terms obtained by observing the runs. We assume that the adversary has unlimited inference capabilities, meaning that he can combine the information in his knowledge to construct or interpret new terms. However, this capability is restricted due to the assumption of perfect cryptography. This means that the adversary cannot reverse hash functions and that he is not able to learn the contents of an encrypted term, unless he knows the decryption key. We denote the inference of term t from term set M by $M \vdash t$. We model corrupted agents by assuming that all secrets of these agents (e.g. secret keys) are contained in the initial knowledge of the adversary. When evaluating security claims, we will only be interested in claims made by trusted (i.e. non-corrupt) agents.

Finally, we assume that the behavior of a collection of agents executing a security protocol is given as a set of traces. Each trace consists of a number of interleaved runs or run prefixes. A run prefix occurs if an agent cannot finish his execution of a role specification (e.g. because the expected input is never provided). We assume that within a trace t the events belonging to run $R\#\theta$ can be identified. Let $t_{R\#\theta}$ denote the subtrace of t consisting of the events of run (or run prefix) $R\#\theta$ which are observable by the adversary. We enumerate all non-empty subtraces $t_{R\#\theta}$ according to the occurrence of their first observable event in trace t. The i-th such subtrace is denoted by t_i^R. The agent executing the events in this subtrace is denoted by $agent(t_i^R)$.

3.2   Untraceability

We define untraceability as a trace property of a role in a protocol. Informally, a role is called untraceable if for every instantiation of the role which is linked to another instantiation of the role, there is a trace that is indistinguishable to the adversary, in which the two instantiations are not linked.

We will first define linkability, reinterpretation, and indistinguishability before presenting the definition of untraceability.

Definition 1 (linkability of subtraces). *Two subtraces t_i^R and t_j^R are linked, denoted by $L(t_i^R, t_j^R)$, if they are instantiated by the same agent:*

$$L(t_i^R, t_j^R) \equiv (agent(t_i^R) = agent(t_j^R)).$$

The notion of reinterpretation has been introduced in [10]. It will be used to express that subterms of a message can be substituted by other terms if the adversary is not able to read (or interpret) these subterms. All terms that the adversary can interpret remain unchanged.

Definition 2 (reinterpretation). *A map π from run terms to run terms is called a reinterpretation under knowledge set M if it and its inverse π^{-1} satisfy the following conditions:*

$$
\begin{aligned}
&\pi(m) = m &&\text{if } m \text{ is a basic run term} \\
&\pi(m) = (\pi(m_1), \ldots, \pi(m_n)) &&\text{if } m = (m_1, \ldots, m_n) \text{ is an } n\text{-tuple} \\
&\pi(\{m\}_k) = \{\pi(m)\}_k &&\text{if } M \vdash k^{-1} \\
& &&\text{or } M \vdash m \wedge M \vdash k \\
&\pi(f(m)) = f(\pi(m)) &&\text{if } M \vdash m \\
& &&\text{or } f \text{ is not a hash function.}
\end{aligned}
$$

Note that the condition $\pi(f(m)) = f(\pi(m))$, when f is not a hash function, leads to an under-approximation of the intended notion of reinterpretation. This means that for certain functions f, there might be untraceable protocols which cannot be proven to be untraceable. In such cases, the condition would need to be refined based on the specific properties of such a function.

Reinterpretations generalize in the obvious way to traces. We use reinterpretations to define indistinguishability of traces. Two traces are indistinguishable to the adversary, if the adversary cannot see any meaningful difference between the two traces, based on the knowledge he has.

Definition 3 (indistinguishability of traces). *Let M be the adversary's knowledge at the end of trace t. The trace t is indistinguishable from a trace t', denoted by $t \sim t'$, if there is a reinterpretation π under M, such that $\pi(t_i^R) = t_i'^R$ for all roles R and subtraces t_i^R.*

We now have all ingredients to formally define untraceability. Untraceability is the property that for every trace of a protocol in which two subtraces are linked, there is a trace that is indistinguishable to the adversary in which these two subtraces are not linked.

Definition 4 (untraceability). *A protocol P is untraceable with respect to role R if*

$$\forall_{t \in Traces(P)} \\ \forall_{i \neq j} L(t_i^R, t_j^R) \Rightarrow \\ \exists_{t' \in Traces(P)} t \sim t' \wedge \neg L(t_i'^R, t_j'^R).$$

4   An Untraceable Protocol

In [12], Feldhofer *et al.* present an AES hardware implementation for RFID tags along with two simple protocols for unilateral and mutual authentication, of which the unilateral authentication protocol can be proven traceable. In this section, we prove that the mutual authentication protocol is untraceable.

4.1   Protocol Description

The protocol assumes that every pair of reader R and tag T shares a unique key $k(R, T)$. These shared keys are initially not part of the adversary's knowledge. The reader initiates the protocol by sending a freshly generated nonce nr to the tag. The tag generates a nonce nt encrypts the pair (nr, nt) under the shared key $k(R, T)$, and sends it to the reader. The reader decrypts the message using the same shared key, reverses the order of the two nonces, encrypts the message under the shared key, and sends it to the tag. Figure 1 shows a graphical representation of the protocol specification.

4.2   Untraceability

Theorem 1. *The protocol depicted in Figure 1 is untraceable.*

Proof. We notice first that $k(R, T)$ and nt remain secret throughout the protocol execution. This can be easily verified by hand or with an automated tool.

Let t be a trace with subtraces t_i^T and t_j^T for $i \neq j$. We need to show that whenever $L(t_i^T, t_j^T)$ we can find a trace $t' \sim t$ such that $\neg L(t_i'^T, t_j'^T)$. For ease of notation, we set $agent(t_i^T) = agent(t_j^T) = agent(t_i'^T) = Alice$ and $agent(t_j'^T) =$

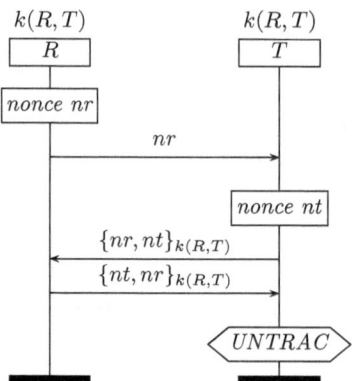

Fig. 1. An untraceable mutual authentication protocol

Bob. The general idea of the proof is that t' can be constructed from t by replacing all occurrences of *Alice* in t_j^T by *Bob*. We will make this more precise below and motivate that the adversary cannot distinguish between t and t'.

Since we are verifying the untraceability claim for an agent in role T, we may assume that the agent is trusted, i.e. that it executes all read and send events according to the specification. By definition, there is a θ such that the subtrace t_j^T contains the event where $\{nr\#\theta, nt\#\theta\}_{k(R\#\theta, T\#\theta)}$ is sent.

We consider the map π with the following properties:

$$\pi(\{x, nt\#\theta\}_{k(y, Alice)}) = \{x, nt\#\theta\}_{k(y, Bob)} \qquad \text{for any } x \text{ and } y,$$
$$\pi(\{nt\#\theta, x\}_{k(y, Alice)}) = \{nt\#\theta, x\}_{k(y, Bob)} \qquad \text{for any } x \text{ and } y,$$
$$\pi(m) = m \qquad \text{elsewhere.}$$

Note that π is a reinterpretation under the adversary's knowledge, by Definition 2 and secrecy of $k(R, T)$.

Let $t' = \pi(t)$. We show that t' is a valid trace. The only difference between the traces t and t' occurs in messages containing the nonce $nt\#\theta$. By construction, the changes produce a valid run for Bob while keeping the reader's run valid. It follows from the secrecy of nt and $k(R, T)$ that any further occurrence of $nt\#\theta$ must be in $\{nr\#\theta, nt\#\theta\}_{k(R\#\theta, T\#\theta)}$ or $\{nt\#\theta, nr\#\theta\}_{k(R\#\theta, T\#\theta)}$. Since $nr\#\theta$ is produced by $R\#\theta$, no other run of R will accept the former message, and similarly, since $nt\#\theta$ is produced by $T\#\theta$, no other run of T will accept the latter message.

Finally, $t_i^T = t_i'^T$ thus $\neg L(t_i'^T, t_j'^T)$. □

5   A Traceable Protocol

Di Pietro and Molva describe in [13] an identification and authentication protocol aimed at enhancing the security and privacy of RFID-based systems. We will first describe the Di Pietro-Molva protocol and then prove it untraceable for a *restricted choice of parameters* and the assumption that Di Pietro and Molva's *DPM* function is a *perfect hash function*. By lifting the restrictions and analyzing the algebraic properties of the *DPM* function we will demonstrate an efficient method to trace tags and discuss its practicality. Finally, we will relate the insight back to our definition of untraceability by exhibiting a trace of the Di Pietro-Molva protocol for which there is no valid, to the adversary indistinguishable, trace with unlinked subtraces.

5.1   Protocol Description

Let h be a cryptographic hash function, M, the majority function of three bits, defined by

$$M : \mathbb{F}_2^3 \to \mathbb{F}_2$$
$$(a, b, c) \mapsto ac + bc + ab$$

and for $\ell \in 3\mathbb{N}$,

$$DPM : \mathbb{F}_2^\ell \to \mathbb{F}_2$$
$$(x_1, \ldots, x_\ell) \mapsto \sum_{i=1}^{\ell/3} M(x_{3i-2}, x_{3i-1}, x_{3i}).$$

It is easy to verify that the functions M and DPM are identical to the corresponding functions in [13], except that we have defined them over vector spaces over the finite field with two elements instead of bit strings. In the remainder of this section we will identify elements in vector spaces over $\mathbb{F}_2$ with bit strings. We will denote the tags' and readers' unique ids by ID_T and ID_R, respectively. Every tag has a unique key k_T assigned to it by the key distribution center (KDC). A reader authorized to identify a tag T will be given by the KDC the key $k_{T,R} = h(k_T, ID_R, k_T)$. The keys are ℓ bits long.

The Di Pietro-Molva protocol, depicted in Figure 2, begins with the reader sending its ID and a random nonce n_j to the tag. The tag replies with the message $\alpha_1, \ldots, \alpha_q, V, \omega$, where $\alpha_i = k_{T,R} \oplus r_i$ for randomly chosen r_i (an ℓ-bit vector), the i-th bit of V (a q-bit vector) is $DPM(r_i)$, and $\omega = h(k_{T,R}, n_j, r_1, k_{T,R})$. The reader has a database of keys $k_{T,R}$. The reader can find the key $k_{T,R}$ with the help of the vectors α_i and values $DPM(r_i)$ by iterating through all possible keys. It is expected that each α_i reduces the number of possible keys by approximately one half. ω can be used to uniquely identify the correct key. The last message of the protocol allows the tag to authenticate the reader.

5.2   Untraceability Under Ideal Assumptions

We show that under the following assumptions the Di Pietro-Molva protocol indeed has the untraceability property with respect to RFID tags. We assume

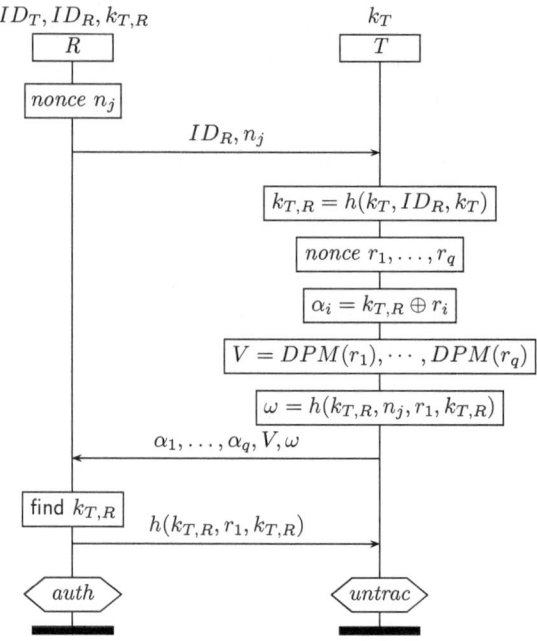

Fig. 2. Di Pietro-Molva protocol

that the random numbers used in the protocol are perfect nonces, we restrict the number q of nonces used by the tag in the protocol to one, and we treat the DPM function appearing in V as a perfect hash function.

Theorem 2. *If the DPM function is a perfect hash function then the protocol depicted in Figure 2 is untraceable for $q = 1$.*

Proof. It can be easily verified with automated tools that $k_{T,R}$ and r_1 are secret. Let t be a trace with subtraces t_i^T and t_j^T for $i \neq j$. We proceed similarly to the proof of Theorem 1, and set $agent(t_i^T) = agent(t_j^T) = agent(t_i'^T) = Alice$ and $agent(t_j'^T) = Bob$.

Let t_j^T contain the event where $\alpha_1 \# \theta, V \# \theta, \omega \# \theta$ is sent. We consider the map π which for any term R has the following properties

$$\pi(h(k_{Alice,R}, n_j \# \theta, r_1 \# \theta, k_{Alice,R})) = h(k_{Bob,R}, n_j \# \theta, r_1 \# \theta, k_{Bob,R})$$
$$\pi(h(k_{Alice,R}, r_1 \# \theta, k_{Alice,R})) \quad = h(k_{Bob,R}, r_1 \# \theta, k_{Bob,R})$$
$$\pi(h(k_{Alice}, ID_R, k_{Alice})) \quad = h(k_{Bob}, ID_R, k_{Bob})$$

and is equal to the identity map everywhere else. Note that according to the specification, $k_{Alice,R} = h(k_{Alice}, ID_R, k_{Alice})$ and $\alpha_1 \# \theta = k_{Alice,R} \oplus r_1 \# \theta$. By the definition of reinterpretation, $\pi(k_{Alice,R} \oplus r_1 \# \theta) = \pi(k_{Alice,R}) \oplus \pi(r_1 \# \theta) = k_{Bob,R} \oplus r_1 \# \theta$. For convenience, we set $\alpha_1' \# \theta = k_{Bob,R} \oplus r_1 \# \theta$.

Let $t' = \pi(t)$. It follows from the construction that the map produces a valid run for Bob while keeping the reader's run valid. The only differences between the traces t and t' occur in messages containing the hashes $h(k_{Bob,R}, r_1\#\theta, k_{Bob,R})$ and $h(k_{Bob,R}, n_j\#\theta, r_1\#\theta, k_{Bob,R})$. Aside from Bob's run, the hashes and $\alpha_1\#\theta$ may be replayed by the adversary. Because $r_1\#\theta$ is generated by $T\#\theta$, no other run of T will accept $h(k_{Bob,R}, r_1\#\theta, k_{Bob,R})$ or $h(k_{Alice,R}, r_1\#\theta, k_{Alice,R})$. Similarly, since $k_{T,R}$ is secret and at most one run of R could have generated $n_j\#\theta$, at most one run of R accepts $\alpha_1\#\theta$, $\alpha_1'\#\theta$, $h(k_{Bob,R}, n_j\#\theta, r_1\#\theta, k_{Bob,R})$, and $h(k_{Alice,R}, n_j\#\theta, r_1\#\theta, k_{Alice,R})$.

Finally, $t_i^T = t_i'^T$, since $r_1\#\theta$ is generated by $T\#\theta$, thus $\neg L(t_i'^T, t_j'^T)$.

It remains to show that π is a reinterpretation under the adversary's knowledge. This follows from the fact that $r_1\#\theta$ is secret. □

Note that the assumption $q > 1$ would invalidate the untraceability proof, because t' would not necessarily be a valid trace anymore. In fact, for $q > 1$ an adversary may be able to determine that a tag is *not* identical to a previously observed tag. This insight can be exploited with an active as well as a passive attack. In an active attack, for each consecutive bit-triplet in α_2, the adversary would change one bit, during one execution. In such a case, the reader replies to the tag with a third message if and only if the two unchanged bits of the corresponding bit-triplet of the nonce r_2 are the same. Such an attack would, after several iterations, lead to the same information as the passive attack demonstrated in the following section.

5.3   Analysis of the *DPM* Function

We consider how much information about the tag is leaked through the *DPM* function and the resulting relation between α_i and $V[i]$. We first observe that for $(a, b, c), (x, y, z) \in \mathbb{F}_2^3$,

$$M(a + x, b, c) = ac + bc + ab + cx + bx$$

with analogous equations for $M(a, b+y, c)$ and $M(a, b, c+z)$. Furthermore, we have

$$M(a+x, b+y, c+z) = M(a+x, b, c) + M(a, b+y, c) + M(a, b, c+z) + M(x, y, z).$$

It follows that

$$M(a + x, b + y, c + z) = M(a, b, c) + M(x, y, z) + a(y + z) + b(x + z) + c(x + y)$$

which after reordering we write as

$$(y + z, x + z, x + y) \cdot \begin{pmatrix} a \\ b \\ c \end{pmatrix} = M(a+x, b+y, c+z) + M(a, b, c) + M(x, y, z). \quad (1)$$

We define, for convenience, the function

$$\text{cross} : \mathbb{F}_2^\ell \to \mathbb{F}_2^\ell$$
$$(x_1, y_1, z_1, \ldots, x_{\ell/3}, y_{\ell/3}, z_{\ell/3}) \mapsto (y_1 + z_1, x_1 + z_1, x_1 + y_1, \ldots,$$
$$y_{\ell/3} + z_{\ell/3}, x_{\ell/3} + z_{\ell/3}, x_{\ell/3} + y_{\ell/3}).$$

Note that $\text{cross}(r_1) + \text{cross}(r_2) = \text{cross}(r_1 + r_2) = \text{cross}(\alpha_1 + \alpha_2)$.

From equation (1) and the definition of $DPM(\cdot)$ we obtain the following identity in which the left-hand side is a product between the row vector $\text{cross}(r_1)$ and $k_{T,R}$ written as a column vector $k_{T,R}^T$.

$$\text{cross}(r_1) \cdot k_{T,R}^T = DPM(k_{T,R} + r_1) + DPM(k_{T,R}) + DPM(r_1) \qquad (2)$$

and similarly

$$\text{cross}(r_2) \cdot k_{T,R}^T = DPM(k_{T,R} + r_2) + DPM(k_{T,R}) + DPM(r_2). \qquad (3)$$

By adding up equations (2) and (3) we obtain

$$\text{cross}(\alpha_1 + \alpha_2) \cdot k_{T,R}^T = DPM(\alpha_1) + DPM(\alpha_2) + DPM(r_1) + DPM(r_2).$$

For $i = 2, \ldots, \ell + 1$, let the $\ell \times \ell$ matrix A be given by the row vectors $\text{cross}(\alpha_1 + \alpha_i)$ and let the column vector v be given by the entries $DPM(\alpha_1) + DPM(\alpha_i) + DPM(r_1) + DPM(r_i)$. Consider then the linear equation $Ax = v$, viz.

$$\begin{pmatrix} \text{cross}(\alpha_1 + \alpha_2) \\ \text{cross}(\alpha_1 + \alpha_3) \\ \vdots \\ \text{cross}(\alpha_1 + \alpha_\ell) \end{pmatrix} \cdot x = \begin{pmatrix} DPM(\alpha_1) + DPM(\alpha_2) + DPM(r_1) + DPM(r_2) \\ DPM(\alpha_1) + DPM(\alpha_3) + DPM(r_1) + DPM(r_3) \\ \vdots \\ DPM(\alpha_1) + DPM(\alpha_\ell) + DPM(r_1) + DPM(r_\ell) \end{pmatrix}$$

By construction, the vector $x = k_{T,R}^T$ is a solution of the equation and so is any vector of the form $k_{T,R}^T + y$, where y is in the null space of A. Thus, the null space of A in this equation can be considered the adversary's uncertainty about $k_{T,R}$. From the definition of the $\text{cross}(\cdot)$ function, it is easy to see that the null space of A contains the vectors

$$(1, 1, 1, 0, \ldots, 0)^T, (0, 0, 0, 1, 1, 1, 0, \ldots, 0)^T, \ldots, (0, \ldots, 0, 1, 1, 1)^T. \qquad (4)$$

The following theorem states that the null space of A is actually spanned by these vectors whenever A is constructed from linearly independent vectors $\alpha_1, \ldots, \alpha_{\ell+1}$. Thus, the adversary can learn all bits of $k_{T,R}$ modulo the vectors in (4), that is, up to complements of $\ell/3$ consecutive bit-triplets.

Theorem 3. *If $\alpha_1, \ldots, \alpha_{\ell+1}$ are linearly independent, then the rank of A is $\frac{2}{3}\ell$.*

Proof. We know that the $\ell/3$ vectors listed in (4) are in the null space of A. Since they are linearly independent, the rank of A is at most $\frac{2}{3}\ell$.

Conversely, consider the matrix $\tilde{A}$ obtained from A by deleting every third column of A. $\tilde{A}$ can also be obtained from the matrix B consisting of the rows $\alpha_1 + \alpha_2, \ldots, \alpha_1 + \alpha_{\ell+1}$ as follows. We add every third column to the preceding two columns and swap those preceding two columns. We call the resulting matrix $\tilde{B}$. Clearly B and $\tilde{B}$ have the same rank. By deleting every third column of $\tilde{B}$, we obtain $\tilde{A}$. Since deletion of a column decreases the rank of the matrix by at most one and $\tilde{B}$ had full rank, it follows that the rank of $\tilde{A}$ is at least $\frac{2}{3}\ell$ and thus the rank of A is at least $\frac{2}{3}\ell$. $\square$

5.4   Practical Considerations

The probability of a random $(n+1) \times n$ matrix over $\mathbb{F}_2$ to have rank n is greater than $1/2$. This follows from a simple computation along the lines of equation (1) in [44]. So we may over-estimate the expected number of random vectors needed to obtain ℓ linearly independent vectors to be 2ℓ. Hence after roughly $2\ell/q$ communications between an adversary and a tag, the adversary is able to compute a secret key of the tag up to complements of consecutive bit-triplets. We will now show that this information is very likely to distinguish one tag from almost all of the other tags in the system.

It follows from Theorem 3 that for each of the 2^ℓ possible secret keys, there are $2^{\ell/3}$ possible keys which cannot be distinguished from it solely based on the information contained in $\alpha_1, \ldots, \alpha_q$ and V. We may assume that the entries of the secret keys are uniformly randomly distributed since they are obtained by applying a cryptographic hash function. If we further assume that the number of tags ν in the system is small compared to 2^ℓ, then the probability that for a given tag, there are one or more tags indistinguishable by the above method is approximately $1 - (1 - \frac{1}{2^{2\ell/3}})^\nu$. If, as suggested by the authors, we use the values $\ell = 117$, $q = 2\log\nu$ and assume that there are $\nu = 2^{16}$ tags in the system, then the probability to find one or more tags which would be indistinguishable from a given tag is approximately $2.17 \cdot 10^{-19}$ and the number of communications necessary with the tag to be able to distinguish it with that probability would be 10. In fact, even the probability that there are two or more indistinguishable tags among 2^{16} tags is vanishingly small, namely $7.1 \cdot 10^{-13}$.

Finally, note that the same method reduces the complexity of computing the secret key of a tag to a brute force search of a space with $2^{\ell/3}$ elements, which for $\ell = 117$ is feasible.

5.5   Traceability

In this section we show that the Di Pietro-Molva protocol without idealizing assumptions on the *DPM* function is traceable by our definition.

We say that the lookup process is *efficient* if any authorized reader can uniquely identify a tag based on $\alpha_1, \ldots, \alpha_q$ and V.

Theorem 4. *Assuming that the lookup process is efficient, the protocol depicted in Figure 2 is traceable.*

Proof. Let t be a trace in which a reader Ray interacts twice with the same tag $Alice$. Let t_1^T and t_2^T be the two subtraces containing the send event of the tag, i.e. $agent(t_1^T) = agent(t_2^T) = Alice$. We need to show that there is no valid trace $t' \sim t$ such that $\neg L(t_1'^T, t_2'^T)$.

By observing t_1^T, t_2^T the adversary can compute $k_{Alice,Ray}$ up to a null space N_1 as shown in Section 5.3. We may assume that t is such that N_1 is the smallest possible null space shown in (4). Note that no other key $k_{T,Ray}$ is equal to $k_{Alice,Ray} + n$ for any $n \in N_1$ because the lookup process is efficient.

Let t' be any valid trace where $agent(t_1'^T) = agent(t_1^T) = Alice$, $agent(t_2'^T) = Bob$. By construction, we have $\neg L(t_1'^T, t_2'^T)$.

By observing $t_2'^T$, the adversary can compute $k_{Bob,Ray}$ up to a null space N_2 with $N_1 \subseteq N_2$ by minimality of N_1. There are no $n_1 \in N_1$, $n_2 \in N_2$ with $k_{Alice,Ray} + n_1 = k_{Bob,Ray} + n_2$ because the lookup process is efficient and $N_1 \subseteq N_2$.

Therefore the adversary can distinguish t from t'.

6   Conclusion

The main contribution of this paper is a definition of untraceability which can be used in formal verification of RFID protocols. We showed how to apply our definition by proving that the protocol in [12] indeed satisfies untraceability. We also demonstrated a weakness in the published protocol in [13], that we could exploit by using linear algebra. We proved that the protocol does not satisfy our definition of untraceability.

In the future, we would like to refine our notion of untraceability. Under the current definition, for a tag to be untraceable, it suffices to find one other tag which could have been present to produce the same trace. A strengthening of this definition is therefore desirable.

Several other refinements are conceivable. One such refinement concerns a weaker notion of untraceability that allows an adversary to recognize a tag between any two successful communications with a trusted RFID reader. Another refinement could be 'untraceability groups' defining the set of agents from which a particular agent cannot be distinguished. A third, slightly stronger notion of untraceability that should be defined properly is the notion of 'forward untraceability', stating that compromising a tag does not compromise its untraceability in past interactions.

A difficult open problem concerns the condition $\pi(f(m)) = f(\pi(m))$ in the definition of reinterpretation. This condition expresses that the application of the function f can be reinterpreted only to the extent its arguments can be reinterpreted under a given knowledge set M. If f is a cryptographic hash function, we know by the perfect cryptography assumption that $f(m)$ can be freely reinterpreted whenever m is not inferable from M. For other functions, however, the reinterpretation depends on the algebraic properties of f and then

$\pi(f(m)) = f(\pi(m))$ is only an under-approximation. Finding the correct condition for a given function f is, in general, non-trivial.

Finally, we plan to automate the process of verifying or finding attacks on untraceability. This leads to new challenges as can be seen in Section 5.3. Under the perfect cryptography assumption, large parts of the verification can be automated, but even state-of-the-art verification tools still struggle with algebraic operations in security protocols.

Acknowledgments

We are grateful to Hugo Jonker, Jun Pang, and the anonymous reviewers for their valuable comments which helped to improve this work.

References

1. Murray, C.J.: RFID tags: driving toward 5 cents. Design News (April 24, 2006)
2. Hoepman, J.H., Hubbers, E., Jacobs, B., Oostdijk, M., Wichers Schreur, R.: Crossing borders: Security and privacy issues of the European e-passport. In: Yoshiura, H., Sakurai, K., Rannenberg, K., Murayama, Y., Kawamura, S.-i. (eds.) IWSEC 2006. LNCS, vol. 4266, pp. 152–167. Springer, Heidelberg (2006)
3. Yoshida, J.: Euro bank notes to embed RFID chips by 2005. EETimes (December 19, 2001)
4. Gilbert, A.: Major retailers to test 'smart shelves'. CNET (January 8, 2003)
5. O'Conner, M.C.: Gilette fuses RFID with product launch. RFID Journal (March 27, 2006)
6. Wong, F.L., Stajano, F.: Location privacy in Bluetooth. In: Molva, R., Tsudik, G., Westhoff, D. (eds.) ESAS 2005. LNCS, vol. 3813, pp. 176–188. Springer, Heidelberg (2005)
7. Jakobsson, M., Wetzel, S.: Security weaknesses in Bluetooth. In: Naccache, D. (ed.) CT-RSA 2001. LNCS, vol. 2020, pp. 176–191. Springer, Heidelberg (2001)
8. Clark, J.A., Jacob, J.L.: A survey of authentication protocol literature. Technical Report 1.0 (1997)
9. Lowe, G.: Breaking and fixing the Needham-Schroeder public-key protocol using fdr. In: Margaria, T., Steffen, B. (eds.) TACAS 1996. LNCS, vol. 1055, pp. 147–166. Springer, Heidelberg (1996)
10. Garcia, F.D., Hasuo, I., Pieters, W., van Rossum, P.: Provable anonymity. In: FMSE, pp. 63–72 (2005)
11. Mauw, S., Verschuren, J., de Vink, E.: A Formalization of Anonymity and Onion Routing. In: Samarati, P., Ryan, P.Y.A., Gollmann, D., Molva, R. (eds.) ESORICS 2004. LNCS, vol. 3193, pp. 109–124. Springer, Heidelberg (2004)
12. Feldhofer, M., Dominikus, S., Wolkerstorfer, J.: Strong authentication for RFID systems using the AES algorithm. In: Joye, M., Quisquater, J.-J. (eds.) CHES 2004. LNCS, vol. 3156, pp. 357–370. Springer, Heidelberg (2004)
13. Di Pietro, R., Molva, R.: Information confinement, privacy, and security in RFID systems. In: Biskup, J., López, J. (eds.) ESORICS 2007. LNCS, vol. 4734, pp. 187–202. Springer, Heidelberg (2007)

14. Weis, S., Sarma, S., Rivest, R., Engels, D.: Security and Privacy Aspects of Low-Cost Radio Frequency Identification Systems. In: Hutter, D., Müller, G., Stephan, W., Ullmann, M. (eds.) Security in Pervasive Computing. LNCS, vol. 2802, pp. 201–212. Springer, Heidelberg (2004)
15. Saito, J., Ryou, J.C., Sakurai, K.: Enhancing privacy of universal re-encryption scheme for RFID tags. In: Yang, L.T., Guo, M., Gao, G.R., Jha, N.K. (eds.) EUC 2004. LNCS, vol. 3207, pp. 879–890. Springer, Heidelberg (2004)
16. Garfinkel, S., Juels, A., Pappu, R.: RFID privacy: An overview of problems and proposed solutions. In: IEEE Security and Privacy, May-June 2005, vol. 3(3), pp. 34–43 (2005)
17. Juels, A.: RFID security and privacy: A research survey. Manuscript (September 2005)
18. Tan, C.C., Sheng, B., Li, Q.: Severless search and authentication protocols for RFID. In: International Conference on Pervasive Computing and Communications – PerCom 2007, USA, IEEE,, March 2007, IEEE Computer Society Press, New York (2007)
19. Seo, Y., Lee, H., Kim, K.: A scalable and untraceable authentication protocol for RFID. In: Zhou, X., Sokolsky, O., Yan, L., Jung, E.-S., Shao, Z., Mu, Y., Lee, D.C., Kim, D.Y., Jeong, Y.-S., Xu, C.-Z. (eds.) EUC Workshops 2006. LNCS, vol. 4097, pp. 252–261. Springer, Heidelberg (2006)
20. Ohkubo, M., Suzuki, K., Kinoshita, S.: Cryptographic approach to privacy-friendly tags. In: RFID Privacy Workshop, MIT, MA, USA (November 2003)
21. Kang, J., Nyang, D.: RFID Authentication Protocol with Strong Resistance Against Traceability and Denial of Service Attacks. In: Molva, R., Tsudik, G., Westhoff, D. (eds.) ESAS 2005. LNCS, vol. 3813, pp. 164–175. Springer, Heidelberg (2005)
22. Dimitriou, T.: A secure and efficient RFID protocol that could make big brother (partially) obsolete. In: PerCom, pp. 269–275 (2006)
23. Choi, E.Y., Lee, S.M., Lee, D.H.: Efficient RFID Authentication Protocol for Ubiquitous Computing Environment. In: Enokido, T., Yan, L., Xiao, B., Kim, D.Y., Dai, Y.-S., Yang, L.T. (eds.) EUC-WS 2005. LNCS, vol. 3823, pp. 945–954. Springer, Heidelberg (2005)
24. Nguyen Duc, D., Park, J., Lee, H., Kim, K.: Enhancing security of EPCGlobal Gen-2 RFID tag against traceability and cloning. In: Symposium on Cryptography and Information Security, Hiroshima, Japan (January 2006)
25. Piramuthu, S.: On existence proofs for multiple RFID tags. In: IEEE International Conference on Pervasive Services, Workshop on Security, Privacy and Trust in Pervasive and Ubiquitous Computing – SecPerU 2006, Lyon, France, June 2006, IEEE Computer Society Press, Los Alamitos (2006)
26. Peris-Lopez, P., Hernandez-Castro, J.C., Estevez-Tapiador, J., Ribagorda, A.: RFID Systems: A Survey on Security Threats and Proposed Solutions. In: Cuenca, P., Orozco-Barbosa, L. (eds.) PWC 2006. LNCS, vol. 4217, pp. 159–170. Springer, Heidelberg (2006)
27. Peris-Lopez, P., Hernandez-Castro, J.C., Estevez-Tapiador, J., Ribagorda, A.: Cryptanalysis of a novel authentication protocol conforming to epc-c1g2 standard (2007)
28. Martinez, S., Magda, V., Concepcio, R., Fransesc, G., Josep, M.: An elliptic curve and zero knowledge based forward secure RFID protocol (2007)
29. Alomair, B., Lazos, L., Poovendran, R.: Passive attacks on a class of authentication protocols for RFID. In: Nam, K.-H., Rhee, G. (eds.) ICISC 2007. LNCS, vol. 4817, pp. 102–115. Springer, Heidelberg (2007)

30. Nohl, K., Evans, D.: Quantifying information leakage in tree-based hash protocols. Technical Report UVA-CS-2006-20, University of Virginia, Department of Computer Science, Charlottesville, Virginia, USA (2006)
31. Tsudik, G.: YA-TRAP: Yet another trivial RFID authentication protocol. In: International Conference on Pervasive Computing and Communications – PerCom 2006, Pisa, Italy, March 2006, IEEE Computer Society Press, Los Alamitos (2006)
32. Ateniese, G., Camenisch, J., de Medeiros, B.: Untraceable RFID tags via insubvertible encryption. In: Conference on Computer and Communications Security – CCS 2005, Alexandria, Virginia, USA, November 2005, ACM Press, New York (2005)
33. Avoine, G.: Adversary model for radio frequency identification. Technical Report LASEC-REPORT-2005-001, Swiss Federal Institute of Technology (EPFL), Security and Cryptography Laboratory (LASEC), Lausanne, Switzerland (September 2005)
34. Juels, A., Weis, S.A.: Defining strong privacy for RFID. In: PerCom Workshops, pp. 342–347 (2007)
35. Chatmon, C., van, L.T., Burmester, M.: Secure anonymous RFID authentication protocols. Technical Report TR-060112, Florida State University, Department of Computer Science, Tallahassee, Florida, USA (2006)
36. Tsudik, G.: A family of dunces: Trivial RFID identification and authentication protocols. Cryptology ePrint Archive, Report 2006/015 (2007)
37. Dimitriou, T.: A lightweight RFID protocol to protect against traceability and cloning attacks. In: Conference on Security and Privacy for Emerging Areas in Communication Networks – SecureComm, Athens, Greece, September 2005, IEEE, Los Alamitos (2005)
38. Lee, S., Asano, T., Kim, K.: RFID mutual authentication scheme based on synchronized secret information. In: Symposium on Cryptography and Information Security, Hiroshima, Japan (January 2006)
39. Steinbrecher, S., Köpsell, S.: Modelling unlinkability. In: Dingledine, R. (ed.) PET 2003. LNCS, vol. 2760, pp. 32–47. Springer, Heidelberg (2003)
40. Huang, D.: On measuring anonymity for wireless mobile ad-hoc networks. In: 31st IEEE Conference on Local Computer Networks, pp. 779–786. IEEE Press, Los Alamitos, CA (2006)
41. Schneider, S., Sidiropoulos, A.: CSP and anonymity. In: Martella, G., Kurth, H., Montolivo, E., Bertino, E. (eds.) ESORICS 1996. LNCS, vol. 1146, pp. 198–218. Springer, Heidelberg (1996)
42. Avoine, G., Oechslin, P.: RFID Traceability: A Multilayer Problem. In: S. Patrick, A., Yung, M. (eds.) FC 2005. LNCS, vol. 3570, pp. 125–140. Springer, Heidelberg (2005)
43. Cremers, C., Mauw, S.: Operational Semantics of Security Protocols. In: Leue, S., Systä, T.J. (eds.) Scenarios: Models, Transformations and Tools. LNCS, vol. 3466, pp. 66–89. Springer, Heidelberg (2005)
44. Cooper, C.: On the rank of random matrices. Random Structures and Algorithms 16(2) (2000)

A Graphical PIN Authentication Mechanism with Applications to Smart Cards and Low-Cost Devices[*]

Luigi Catuogno[1] and Clemente Galdi[2]

[1] Dipartimento di Informatica ed Applicazioni, Università di Salerno
Via Ponte Don Melillo, 84084 Fisciano (SA), Italy
luicat@dia.unisa.it
[2] Dipartimento di Scienze Fisiche, Università di Napoli "Federico II"
Compl. Univ. Monte S.Angelo, Via Cinthia, 80126 Napoli (NA), Italy
clemente.galdi@unina.it

Abstract. Passwords and PINs are still the most deployed authentication mechanisms and their protection is a classical branch of research in computer security. Several password schemes, as well as more sophisticated tokens, algorithms, and protocols, have been proposed during the last years. Some proposals require dedicated devices, such as biometric sensors, whereas others of them have high computational requirements. Graphical passwords are a promising research branch, but implementation of many proposed schemes often requires considerable resources (*e.g.,* data storage, high quality displays) making difficult their usage on small devices, like old fashioned ATM terminals, smart cards and many low-price cellular phones.

In this paper we present a graphical mechanism that handles authentication by means of a numerical PIN, that users have to type on the basis of a secret sequence of objects and a graphical challenge. The proposed scheme can be instantiated in a way to require low computation capabilities, making it also suitable for small devices with limited resources. We prove that our scheme is effective against "shoulder surfing" attacks.

1 Introduction

Passwords and PINs are still the most deployed authentication mechanism, although they suffer of relevant and well known weakness [3]. The protection of passwords is a classical branch of research in computer security. Several important improvements to the old-fashioned alphanumeric passwords, according to the context of different applications, have been proposed in the last years. Indeed, literature on authentication and passwords is huge, here we just cite Kerberos [4], S/Key [5] and OPIE [6].

---

[*] This work was partially supported by the European Union under IST FET Small/medium-scale focused research project FRONTS (Contract n. 215270).

J.A. Onieva et al. (Eds.): WISTP 2008, LNCS 5019, pp. 16–35, 2008.

Two important aspects in dealing with passwords are the following:

1. Passwords should be easy enough to be remembered but strong enough in order to avoid guessing attacks;
2. The authentication mechanism should be resilient against classical threats, like shoulder surfing attacks, i.e., the capability of recording the interaction of the user and the terminal; moreover, it should be light enough to be used also on small computers.

In the following, we describe what we call the *ATM Scenario* where the need for an authentication mechanism satisfying the above requirements becomes critical.

In the ATM Scenario the user uses a magnetic strip card to access ATM terminals. In order to be authenticated, the user pushes her card (that carries only her identification data) in the ATM reader and types her four digit PIN; afterwards, the ATM sends the user's credentials to the remote authentication server through a PSTN network. This approach is daily used by thousand of users, nevertheless it suffers from some well-known vulnerabilities. Magnetic strip cards can be easily cloned and, PIN numbers can be collected in many ways. For example, an adversary could have placed a hidden micro-camera pointing to the ATM panel somewhere in the neighborhood. A recent tampering technique is accomplished by means of a *skimmer*, i.e., a reader equipped with an EPROM memory that is glued upon the ATM reader, so that strips of passing cards can be dumped to the EPROM. A forged spotlight is also placed upon the keyboard in order to record the insertion of the PIN. The skimmer allows adversaries to collect an undefined number of user sessions obtaining all information needed to clone user cards.

Graphical passwords are a promising authentication mechanism that faces many drawbacks of old-style password/PIN based scheme. The basic idea is to ask the user to click on some predefined parts of an image displayed on the screen by the system, according to a certain sequence. Such a method has been improved during the last years, in order to obtain schemes offering enhanced security. However, the majority of proposed schemes require costly hardware (*e.g.*, medium/high resolution displays and graphic adapters, touch screen, data storage, high computational resources etc.). This makes some of the proposed schemes not suitable to be implemented on low cost equipments (*e.g.*, current ATM terminals that are still the overwhelming majority).

In this paper we propose a graphical PIN scheme based on the challenge-response paradigm that is effective to prevent "shoulder surfing" attacks. Our scheme could replace the classical PIN authentication mechanism in the two scenarios described above. The design of the scheme follows three important guidelines:

− The scheme should be *independent* from the specific set of objects that are used for the graphical challenge. In particular, our scheme can be deployed also on terminals that are equipped with small sized or cheap displays like the ones of the cellular phones, or through the classical (color and monochrome) 10 inch

CRT monitor that still equips thousands of ATM terminals. Moreover, user responses should be composed as well by any sophisticated pointing device as by simple keypad.

- The generation of challenges and the verification of user's responses should be *affordable* also by computer with limited computational resources (*e.g.* smart cards, security tokens).
- The user is simply required to *recognize* the position of some objects on the screen. She is *not required to compute* any function.

Our Results. In this paper we assume that the terminal used by the user cannot be tampered. In other words, an adversary is allowed to record the challenges displayed by the terminal and the activity of the user but she is not allowed to alter in any way the behavior of the parties. Furthermore, we assume that a sequence of three unsuccessful authentications leads to the block of the user account. This assumption is extremely common in many application scenaria, e.g., ATM. Furthermore, the adversary does not know when a legitimate user will successfully authenticate (and reset the "failure counter"). We say that an attack is successful if the adversary can "extract the user secret".

We present a strategy that can withstand shoulder surfing attacks. More precisely, in our scheme the challenge issued by the system is a random arrangement of the objects into a matrix displayed on the screen. During her authentication session, the user is required to type as PIN the position of a sequence of secret objects in the challenge matrix. Clearly, the PIN typed in by the user changes in each session as the challenge changes. To be more precise, the queries the user is required to answer are questions like *"On which row of the screen do you see object o?"*. Hence, in order to compute the correct response, the user has to watch the screen and answer some/all the questions corresponding to her secret objects, according to a given protocol.

We have experimentally evaluated the robustness of the proposed schemes against "shoulder surfing" attacks. We first analyze a naive protocol, where the user has to answer correctly to all queries of the challenge, *i.e.*, she has to compose the PIN with the digits representing the correct row number of all objects in her secret sequence.

Then, we describe two different protocols, called *user-randomized protocols*, where the user is allowed to reply the challenge issued by the server with a certain number of random or wrong answers. We show that, these randomized variations increase, w.r.t. to the naive scheme, the number of sessions the adversary needs to collect before being able to successfully extract the user secret. Following the approach presented in [2], it is possible to show a SAT-based attack.

We stress that the set of objects used to construct the challenges has an impact on the usability of the scheme. The objects used to construct the challenge should depend on the application scenario or, even better, on the users' preferences. For example, painters might be more comfortable with paintings than mathematicians that, in turn, might easily identify a sequence of numbers with specific properties. Notice that it might be even possible to use "letters" as objects to be displayed. In this case the graphical password the user needs to remember reduces to a "classical" password.

On the other hand, complex objects cannot be displayed/managed on low-cost devices. Furthermore, the more complex are the images, the harder is the task of automatic classification that, in turn, could help the adversary in attacking the scheme.

Our scheme is independent from the specific set of objects. This makes it is suitable for deployment both on complex and simple devices and tunable on the specific application scenario.

Since our scheme requires a limited computational ability both to the user and the authenticator, following the lead of [7], our scheme could be easily deployed in those contexts where small sized devices with poor computational resources (*e.g.,* pervasive devices) are involved. In particular, our scheme could fit a RFID infrastructure as tag-to-reader and/or reader-to-tag authentication protocol within the Minimalist model defined in [8]. Moreover, our scheme could be used to enforce multi-factor authentication schemes via smartcard as card-to-reader authentication protocol. Note that even on cheapest devices, randomized protocols we present in this work could be implemented chosing set up paramenters beyond the ones affordable by human users.

2   Related Work

Identification of users through insecure channels is a classical problem in the area of computer security. One of the earliest researches on this topic is due to Lamport [9], who proposed a *one-time* password scheme, i.e., an authentication method in which the user has to prove the knowledge of the password instead of providing it. This scheme belongs to the family of *challenge and response* protocols, where the system issues a challenge to the user, who has to compute a given function of the challenge and of the secret password. The system successfully authenticates the user if the provided result is correct. The term *one-time* means that the same password can be used for several authentication rounds, but the response computed by the user is different for each round. Some implementations of the above scheme were proposed in [5,6]. The main drawback of this approach is that the user needs the help of a cryptographic device in order to compute her answer correctly. Several research has been done on defining human computable challenges [10,11,12] and evaluating the security of the resulting protocols [13,14,15].

Graphical passwords constitute a solution in this direction, since, as shown in [16], it is easier for the user to consider images instead of letters and numbers. On the other hand, since password-based identification schemes are very common, user might accept easily a letter-based password scheme in stead of a graphical one. An authentication mechanism using graphical passwords was first proposed by Blonder [17]. In his scheme an image is displayed on the screen and the user is required to click on some previously chosen regions of the image, according to a certain sequence. Images, regions and sequences of clicks are selected at user's registration time. In the *Dèjá vu* scheme [18,19], the user, during the registration phase, is allowed to choose some images from a set of random pictures generated by the system. Later on, in order to be authenticated, the user has to recognize her pre-selected images in the set of images shown by the system. Jansen *et al.* proposed an analogous paradigm in [20,21], whereas, the "Pass-Faces" project by Real User Corp. [22] uses images of human faces instead of generic pictures. In the *Draw a Secret* scheme [23] the user is required to paint a pre-defined two-dimensional picture in the same way she did during the registration phase (that is, drawing lines and points in the same order and in the same coordinates).

Sobrado and Birget[24] proposed a scheme where, during the registration phase, the user chooses a set of small pictures (pass-icons). When the user logs in, the system shows her a screenshot populated by many different icons. In order to be authenticated, the user has to click any icon belonging to the convex-hull whose vertices are the pre-selected pass-icons. This scheme has been improved in [25].

Roth *et al.* [26] focused their attention on handling PINs of magnetic strip cards, where each PIN digit is inserted by the user in several rounds. In each round, the system shows the possible digits randomly partitioned into two sets, whose elements are depicted with a different color (*e.g.* black and white) and the user has to select the color related to the set the current digit belongs to. The intersection of sets selected at every round gives the PIN digit for the user. The security of the scheme against attacks performed by adversaries either with human memorization capabilities or with camera recording capabilities was also discussed in [26].

In the scheme presented in [1], the user and the system share a secret subset $\mathcal{F}$ of a set of public pictures $\mathcal{B}$. The authentication process is composed of several rounds. In each round the system shows to the user a table containing a picture of $\mathcal{B}$ in each cell, in a random order. The user is asked to find, across the table, a path between the image located to the top-left corner of the table and the last column or the last row of the table. The setup of this scheme is quite complicated. Users need to pass a training phase that spans over two days, and the login time can require up to some minutes. In [2] the authors present a simple attacks that breaks the scheme presented in [1]. They used information collected by observing a limited number of queries in building a system of boolean expression. Using a PC running a SAT solver [27], they are able to find the secret under the default parameters reported by [1] in 102 seconds, after collecting just 60 round transcripts.

Recently, in [28] the authors present a system that allows users to enter passwords by using the orientation of their pupils. The users input their password using gaze-based typing. Computer vision techniques are used to track the orientation of the user's pupils and to extract the password. The authors show the time needed to enter and the error rates obtained by using their system is comparable with the ones obtained by using a keyboard. Furthermore, the users tend to prefer the use of gaze-based systems in place of classical password/PIN-entry methods. On the other hand, such scheme requires costly hardware since image analysis has to be executed in an on-line fashion, i.e., while the user is "typing in" the password.

For a wider overview about research on graphical passwords, we suggest the reader to take a look at the survey by Suo *et al.* [29] and visit the web site of the "Graphical Passwords Project"[30] at Rutgers.

3   Preliminaries

In this section we introduce the notation and conventions used in the rest of the paper.

OBJECTS AND CHALLENGES. The protocols described in this paper belong to the family of *challenge and response* authentication schemes, where the system issues a random challenge to the user, who is required to compute a response, according to the challenge and to a secret shared between the user and the system. In particular, the challenges consist of random pictures containing several objects. We denote by O the set of all distinct objects and by $q = ak$, for some positive integers a and k, its cardinality. A *challenge* is a sequence $\alpha = (o_1, \ldots, o_q)$, where o_i is an object drawn from O. The objects in α are arranged in a matrix with a rows and $k = q/a$ columns.

SECRETS DESCRIPTION. In our protocols the secret is a sequence of m objects $\sigma = (\sigma_1, \ldots, \sigma_m)$. The authentication protocol consists of m questions, called *queries*. The i-th query is a question of the following type: *"On which row of the screen do you see the object σ_i?"*. Since questions are chosen independently, the set of possible queries has size $|O|^m$. Since the m objects in the secret are chosen independently, the set of possible secrets has size $|O|^m$.

RESPONSES AND SESSION TRANSCRIPTS. Upon reception of a challenge, the user is required to compute a response, according to the secret queries shared with the system. A response is a vector $\beta = (\rho_1, \ldots, \rho_m)$, where each ρ_i is a number drawn from a set $A = \{0, 1, \ldots a - 1\}$, representing the answer to the i-th query, according to the challenge. A Session Transcript is a pair $\tau = (\alpha, \beta)$, where α is a challenge and β is the user response to α.

4   A Naive Protocol

In this section we describe a first protocol allowing a user U to authenticate herself to a terminal T. We assume U and T share a sequence $\sigma = (\sigma_1, \ldots, \sigma_m)$ of m queries. Furthermore, the user U is provided with a token (*e.g.*, a smart-card), carrying all the information needed to identify U, (*e.g.*, U's account number).

Upon insertion of the token into the terminal, the terminal constructs a challenge α by partitioning the set O of possible objects into a (disjoint) sets $Q_1, \ldots, Q_a$, corresponding to distinct rows displayed on the screen, such that each row contains exactly q/a objects, i.e., $|Q_i| = q/a$, for $i = 1, \ldots, a$. Notice that a denotes the number of possible answers to each query, i.e., the cardinality of the set A.

The introduction of the set A as the set of possible answers is due to the practical idea that users' answers should not be complex to be computed. In other words, in order for the system to be usable, the user should not be forced to search an object in a set with many elements before computing the correct answer.

The use of the set A allows us to restrict the set of possible answers from $\{0, \ldots, 9\}^m$, to $\{1, \ldots, a\}^m$, where $a < 10$. Under this assumption, in order to avoid the possibility for the adversary to randomly guess the array β, the number of queries should be sufficiently large. For example, in order to have an answer space containing at least 10000 elements, the number of digits the user should type, when $a = 4$, is at least 7, since $a^7 = 16834$.

The user U, on input the challenge α is required to compute her response $\beta = (\rho_1, \ldots, \rho_m)$, where ρ_i is the answer to the i-th query "On which row is the object σ_i displayed?". The user passes the authentication test if *all* answers in β are correct.

We note that the authentication consists of a *single* round. The terminal T displays a single challenge and the user replies with m integers drawn from A.

Blind Attack. The first attack we consider to the above protocol is the *blind attack*, where the adversary simply tries to guess the correct answer to a random challenge, without any knowledge of previous authentication transcripts. Clearly, the success probability of such an adversary is $1/a^m$, since there are a^m possible answers, i.e., a answers for each one of the m queries.

The Recording Attack. We now consider the case in which the attacker has the chance to control the terminal T, by recording a certain (finite) number of authentication transcripts from successful sessions carried out by the user U. We also assume that the adversary cannot tamper T. In other words, the adversary can (a) read the information contained on the token; (b) read the challenge issued by the terminal T and (c) read the User reply to the challenge. The adversary cannot actively interfere with the authentication process and, in particular, she can neither (a) modify the challenge presented to the user nor (b) modify the user's answer.

In order to evaluate the robustness of the proposed scheme, we assume that the goal of the adversary is to extract the user secret given a certain number of transcripts[1]. Recall that a sequence of three unsuccessful authentications leads to the block of the user account. For this reason, since the naive protocol authenticates the user only if she correctly replies to *all* the queries in the challenge, we consider the extraction of the secret a necessary condition for the adversary to impersonate the user with probability 1.

We have experimentally evaluated the robustness of the proposed protocol. The simulations we have carried out aim at identifying the *average number* of transcripts that the adversary needs in order to correctly extract the user secret

---

[1] As we will see in the next section, secret extraction is not the only possible goal for the adversary.

as a function of (a) the number of objects q used to construct the queries; (b) the number of rows a used to partition the objects and (c) the number of objects m in the user's secret.

Let $\beta = (\rho_1, \ldots, \rho_m)$. For each object o_j in the set of objects we keep m counters, denoted by $w_{j,1}, \ldots, w_{j,m}$, one for each component in the user secret. An object o_j is said to be a *candidate* for the i-th component of the secret if $w_{j,i} = \max_{o_l \in O} w_{l,i}$. In other words, an object is a candidate for the i-th component if its i-th counter has the maximum value among the counters for the the the specific component.

Since challenges are randomly constructed, the average is computed over 10000 executions of the following experiment:

- A user secret is uniformly selected among the $|O|^m$ possible secrets.
- The adversary requires as many transcripts (α, β), where $\alpha = (Q_1, \ldots, Q_a)$ and $\beta = (\rho_1, \ldots, \rho_m))$, she needs to extract the user secret. The i-th counter associated to object o_j, $w_{j,i}$, is incremented if o_j belongs to the row identified by the answer to the i-th query, i.e., $o_j \in Q_{\rho_i}$.
- The process terminates when *for the first time* there exists, for each component in the secret, exactly one candidate object.

Intuitively, the above process identifies the user secret because each answer to the challenge is always correct. For this reason, after the analysis of k transcripts, the counters associated to each object in the user secret will have value k, i.e., each such object will be a candidate for its component. On the other hand, because of the randomized nature of the challenge creation process, as k grows all the objects that do not belong to the user secret will, eventually, have a counter whose value is strictly less than k.

In Figure 1 we report the average number of transcripts needed to extract the secret when the number of objects in the user secret is $m = 15$, the total number of objects $q \in \{20, 60, 80, 100\}$ and the number of rows used to partition the objects a belongs to $\{2, 4, 5, 10\}$.

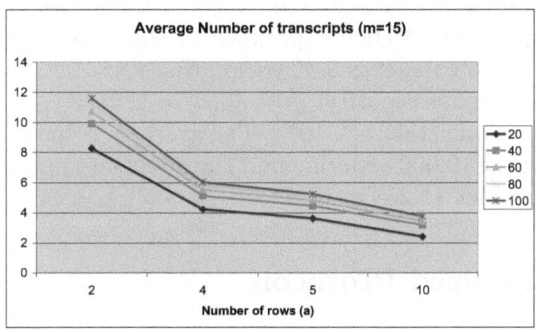

Fig. 1. Strategy Naive: Average number of transcripts needed to extract the secret with $m = 15$

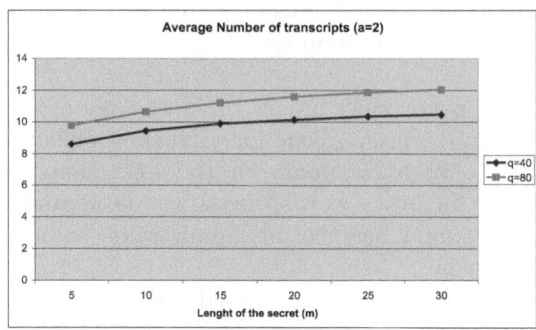

Fig. 2. Strategy Naive: Average number of transcripts needed to extract the secret with variable length secrets

It can be seen that whenever the value of a increases, the average number of required transcript drops quickly. Intuitively, this is due to the fact that the bigger is the value of a, the smaller is the number of objects on each row and, thus, the higher is the information gained by the adversary for each transcript.

In Figure 2 we report the dependence of the average number of required transcripts and the length of the user secret when $a = 2$. As expected, the longer is the secret, the bigger is the number transcripts needed to extract the user secret. However, the number of required transcripts grows slowly w.r.t. the length of the secret. If we consider $q = 40$ and $a = 2$, the average number of transcripts needed to extract a secret containing $m = 10$ objects is slightly less than 10, while the corresponding value for $m = 30$ is slightly higher than 10.

The above discussion shows that, on one hand, the values of m can be low enough to guarantee usability. On the other hand, m cannot be too small in order to prevent blind attacks.

The above experimental evaluation allow to define the following scenaria:

- Small-sized displays, lower security: $q = 40, a = 2, m = 10$. In this case, the probability of a blind attack is $9.7 \cdot 10^{-4}$. The average (resp., minimum) number of transcripts (over 10,000 experiments) the adversary needs to collect before being able to extract is 9.45 (resp., 6).
- Bigger displays, higher security: $q = 80, a = 2, m = 15$. In this case, the probability of a blind attack is $3 \cdot 10^{-5}$. The average (resp., minimum) number of transcripts (over 10,000 experiments) the adversary needs to collect before being able to extract 11.23 (resp., 7).

5   User Randomized Protocols

In this section we explore the possibility that the user herself randomizes the protocol. In other words, the user is allowed to give either random or wrong answers to some randomly chosen queries. It is immediate that the efficiency of these two strategies is different since, intuitively, a random answer does not reveal any information about user's secret while a wrong one does. Furthermore, an adversary always "guesses" a random answer, but it may fail in guessing a "wrong" answer. Thus one basic difference between these two strategies: If the user is allowed to give *wrong* answers (as opposed to *random* ones), we can require as an acceptance criterion that *exactly c* out of m answers are correct *and* that *exactly* $r = m - c$ out of m are wrong (as opposed to *at least c* correct answers out of m.) Clearly, the "correct-random" strategy should be easier to attack.

Notice that user randomization slightly modifies the goal of the adversary. Indeed, in this case, the adversary it is not required anymore to *completely* extract the user secret. An attack is successful if it manages to extract a sequence of objects that can be used for the authentication. The extracted sequence will certainly contain some components of the user secret but it may also contain some objects that do not belong to it.

It is immediate that the success probability of a blind attack for randomized protocols is greater that the corresponding probability for the naive deterministic protocol with the same parameters. For this reason, we have to carefully consider such success probability in order to avoid situations in which it is difficult for the adversary to extract a sequence of objects that allow the authentication but, at the same time, it is easy to be successful using a blind attack.

Although it is well known that, for various reasons, humans are not good random generators, we will still assume that a user can randomly select objects for the following reasons: (a) If users are well-trained and informed about the consequences of their misbehavior, they will actually try to select objects randomly instead of deterministically; (b) our scheme is also applicable for device authentication, i.e., in a non-human context.

5.1   Correct and Random Answers

Let $1 \leq c \leq m$ be an integer. The user randomly selects c out of the m queries and gives correct answers only to these queries while returns random answers to the remaining $r = m - c$ ones. Clearly, if $c = m$ the protocol is the one presented in the previous section.

Blind Attack. Let us consider the success probability of a blind attack. First of all we notice that the maximum number of random answer depends on the value of a. Indeed, let σ be a secret and let $(\alpha, \overline{\beta})$ be a transcript in which all the answers in $\overline{\beta}$ are correct. If the adversary constructs β by randomly picking values in the range $\{1, \ldots, a\}$, the expected number of components in β that will be equal to the corresponding component in $\overline{\beta}$ is m/a. Thus the adversary will be able to correctly guess m/a components of the reply. Since the authentication criterium is "β contains at least c correct answers", if we let $r > m/a$, the adversary will be able to successfully authenticate w.h.p. For this reason, we will only consider values for r that are strictly less than m/a.

Algorithm Authenticate(O, a, c)

1. T constructs a challenge α by randomly partitioning O into a sets $Q_1, \ldots, Q_a$ such that $|Q_i| = q/a$, for $i = 1, \ldots, a$, and displays it on the screen.
2. U computes her response $\beta = (\beta_1, \ldots, \beta_m)$ by correctly answering to c queries and giving random answers to the remaining ones. U sends β back to T.
3. T authenticates U if *at least* c answers are correct.

Fig. 3. An improved authentication protocol

Since the user is required to correctly answer c queries, while she is allowed to give *random* answers to the remaining $r = m - c$ queries, the success probability of a blind attack in this case is $\sum_{h=c}^{m} \binom{m}{h} 1/a^h (1 - 1/a)^{m-h}$.

The Recording Attack. Recall that the goal of the adversary is to obtain a sequence of objects that can be used for successfully authenticate to the terminal. Thus, if the authentication protocol allows the user to reply using $r = m - c$ out of m random answers, it is enough that the adversary manages to correctly extract *at least* c components of the secret. Such set of objects is enough to fulfill her goal.

Notice that the strategy used to extract the user secret presented in the previous section does not work with the randomized authentication strategy. Indeed, in the randomized case, the extraction process cannot stop "the first time there exists, for each component in the secret, exactly one candidate". Intuitively, since user answers are randomized, if at a certain time there exists a single candidate for a given component, such candidate might be different from the actual component in the secret.

For this reason, we have slightly modified the attack strategy. Instead of allowing the adversary to obtain as many transcripts she needs, we provide her t transcripts $(\alpha_1, \beta_1), \ldots, (\alpha_t, \beta_t)$. As in the previous case, the adversary counts the number of times each object belongs to the row identified by the user answers. After t transcripts it may be the case that for some components the adversary has identified more than one candidate, i.e., there exist at least 2 objects whose counter for the specific component has the maximum value. In this case we randomly pick one of these objects as actual candidate. If, instead, for each component there exists exactly one candidate, the following cases may arise:

- All candidates are correct. The adversary has correctly extracted the whole user secret. We call such sequences of objects *good*.
- The number of correct candidates belongs to $\{c, \ldots, m - 1\}$. The user secret has *not* been correctly extracted but the sequence of objects is a valid authentication secret. We call such sequences of objects *valid*.
- The number of correct candidates is strictly less than c. We call such sequences of objects *wrong*.

We assume that the adversary is successful if she manages to extract either a good or a valid secret.

We have first analyzed the dependence of the *sum* of the number of good and valid sequences w.r.t. the number of random answers allowed by the scheme when (a) q belongs to the set $\{20, 40, 60, 80, 100\}$, (b) $a = 2$, i.e., the q objects are partitioned into two sets, (c) $m = 15$, i.e., user secret consists of 15 objects and (d) the adversary is provided with $t = 15$ transcripts. Similar results can be obtained using different parameters. Since, as stated above, the maximum number r of random answer has to be strictly less than $m/2$, we consider the case in which r belongs to the set $\{0, \ldots, m/3 = 5\}$.

From our experiments we can derive that, even if the number transcripts provided to the adversary is "high"[2], as the number of random answers increases, the number of good (resp., good and valid) secrets decreases quickly. Furthermore, in some cases, the adversary is not even able to extract a valid secret out of the given transcripts.

At this point we have considered the case in which q is fixed to 80 and we let the value of a to belong to the set $\{2, 4, 8, 10\}$. Notice that, since a is not constant, also the maximum number of random answers varies depending on a. Also in this case the length of the secret consists of $m = 15$ objects and we have provided the adversary with $t = 15$ transcripts.

From the results of the experiments we can derive that the adversary's probability of extracting good secret increases very quickly as the value of a increases. On the other hand, if we consider both good and valid secrets, the probability of success of the adversary is extremely high. For the specific set of parameters, the adversary may fail in extracting a good or a valid secret only if $a = 2$.

Finally we have evaluated the success probability of the adversary when the number of transcripts t provided increases from 10 to 30. As expected, as the number of transcripts given to the adversary increases, the probability of extracting a good or a valid secret increases. Notice that if the number of random answers allowed by the scheme increases, the success probability of the adversary decreases. Unfortunately if we set $r = m/3$, the success probability p of a blind attack becomes high, ($p = 0.15$). If we set $r < m/3$, e.g., $r = 4$ in our example, the probability of success of a blind attack decreases to 0.059 while the authentication scheme is still resilient to an adversary that can collect up to 15 transcripts without being able to extract neither a good nor a valid sequence.

5.2   Correct and Wrong Answers

In the previous section we have analyzed the case in which the user has to give correct and random answers. We now consider the case in which the user can alternate correct and *wrong* answer. As stated above, we assume that the user is required to answer to each query with *exactly* c out of m correct answers and *exactly* $r = m - c$ out of m wrong ones.

---

[2] Recall that the average number of transcripts needed to correctly extract the user secret with $q = 80$, $a = 2$ and $m = 15$ using the Naive strategy is slightly higher that 10.

Blind Attack. In this case, if the user is required to answer c correct queries and give wrong answers to the remaining $r = m - c$, the success probability of a blind attack is $\binom{m}{c}(1/a)^c(1 - 1/a)^{m-c}$.

Recall that in the Correct-random strategy, the number of random answers cannot be to high. Indeed, if for example $r = m$, the adversary has probability 1 of being successful in a blind attack.

For this strategy, such limitation does not apply. Indeed the adversary needs to guess *exactly* c correct answers out of m as opposed to *at least* c for the correct-random case. For this reason the value of c can range from 0 to m. Clearly the success probability of a blind attack is maximized when $c = m/2$. However, in this case, such probability is never equal to one.

The Recording Attack. We have experimentally verified this strategy using the same approach we have used the same approach described in Section 5.1.

We have first analyzed the case in which q belongs to the set $\{20, 40, 60, 80, 100\}$, $a = 2$, i.e., the q objects are partitioned into two sets, $m = 15$, i.e., user secret consists of 15 objects. Similar results are obtained with different sets of parameters. For such experiments, and the adversary is provided with $t = 40$ transcripts. The number r of wrong answers required to the user ranges in $\{0, 1, \ldots, 7\}$. Surprisingly, the behavior of the success probability does not strongly depend on the number of objects.

Figure 4 show the percentage of success of the adversary in extracting good or valid secrets from the transcripts. The different curves, each describing a different value of q, are very close to each other.

We have then analyzed the case in which q is fixed to 80 and the value of a belongs to the set $\{2, 4, 8, 10\}$ while keeping the values and ranges of the remaining parameters as in the previous set of experiments. Figure 5 reports the percentage of success when the value of q is fixed to 80 and $m = 15$. In this case, as expected, the lower is the value of a, the lower is the percentage of success of the adversary.

We have also analyzed the dependence of the success probability of the adversary w.r.t. the number of transcripts provided. As expected, the higher is the number of transcript, the higher is the success probability of the adversary. Furthermore, as the number of wrong answers required by the scheme grows from 0 to $m/2$, the the adversary's probability of success decreases.

The case $a = 2$ has an interesting property. Assume that the number of errors required by the scheme is $m/a = m/2$. We can restate the previous statement as "for each i, the user answers correctly i-th query with probability $1/2$". In our setting, this implies that the counter associated to the i-th object of the user secret is incremented, at each transcript, with probability $1/2$. Now notice that this is (approximately) the same probability with which the the i-th counter of any other object is incremented.

This means that the frequencies with which the objects are selected by the user are more or less the same and, thus, the user secret cannot be identified by using the counters approach. The impossibility of using the attack technique described so far is due to the fact that counters associated to each object only

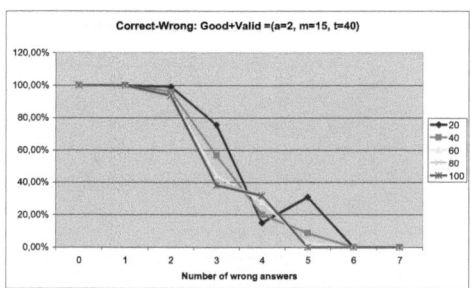

Fig. 4. Strategy Correct-Wrong. The adversary is provided with 40 transcripts. Percentage of good and valid secrets extracted as function of the number of wrong answers with $a = 2, m = 15$ and different values of q.

consider the occurrence of each object independently for each component of the secret. In other words, the attack strategy does not consider the fact that in each transcript there are *exactly* c correct answers and $m - c$ wrong ones. As the number of wrong answers approaches to $m/2$, the number of transcripts needed to extract either a good or a valid secret increases. If such number is approximately $m/2$, an attack that uses a counting argument cannot extract neither a good nor a valid secret, even if the adversary is provided with an extremely high number of transcripts. Such arguments are supported by the results of the experiments. Indeed, Figure 6 (resp., Figure 7) shows that when the number of wrong answers is approximately $m/2$, even if the adversary is provided with 100 transcripts, she cannot even extract a good (resp., a valid) sequence of objects.

Unfortunately the value $r = m/2$ cannot be used in practice since the success probability of a blind attack in this case is high. For example if $m = 15, a = 2$ and $r = 8$, the probability of a blind attack is equal to 0.19. If we reduce the value of r to 5, the probability of success of a blind attack decreases to 0.09. Since we assume that three unsuccessful authentication trials lead to the block of the user account, such set of parameters may be satisfactory in some application scenaria. On the other hand, the latter set of parameters is resilient to an adversary that allows the adversary to collet up to 36 transcripts.

5.3   Possible Extensions and a SAT-Based Attack

The authentication strategies presented in this paper guarantee that the adversary cannot extract a good or a valid sequence of object given a certain number of transcripts.

In the "Correct-Wrong" strategy the number of required transcript can be as high as 36. We have argued and experimentally verified that if the number of answers the user is allowed to give to each challenge may increase to $m/2$, the adversary might not be able to extract a valid transcript using the attack

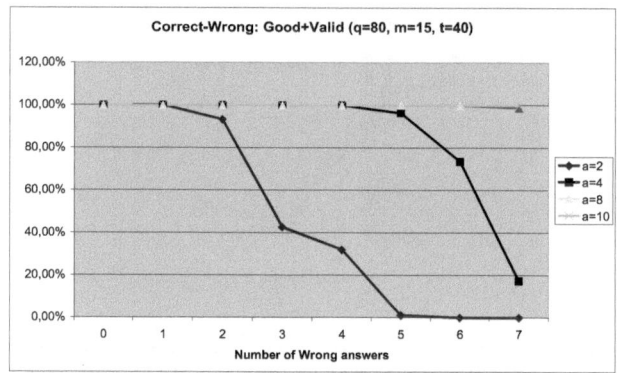

Fig. 5. Strategy Correct-Wrong. The adversary is provided with 40 transcripts. Percentage of good and valid secrets extracted as function of the number of wrong answers with $q = 80, m = 15$ for different value of a.

strategy presented so far. Unfortunately such parameter setting cannot be used because of the high probability of success for a blind attack.

On the other hand we may require the user to answer correctly to a *specific* set of answers (instead of *any set containing exactly c correct answers*). Clearly, the required set of correct answer need to change for each challenge, otherwise the adversary will immediately identify the components of the user secret that always correspond to the correct answers. In this case we have the following side effects:

- The success probability of a blind attack decreases to $(1/a)^c(1 - 1/a)^{m-c} = 1/2^m$;
- The length of the user secret decreases; In this case, the length of the secret can be safely decreased to 10.
- The user needs to remember the specific set of objects to which she has to answer correctly. Clearly this makes the user secret longer. We may circumvent this problem by providing the user with a specific hardware device that provides, at each authentication, a different set of answers to which the user has to answer correctly. We notice that such tokens are already used for providing one-time PINs. However, we notice that if the token is used to provide the one-time PIN "in clear", an adversary that steals the token can easily impersonate the legitimate user. In our case, the mere possession of the device still does not allow the adversary to authenticate without the knowledge of the user secret. Thus the user secret still plays a central role in the multi-modal authentication scheme. We stress that in a recording attack, the adversary is not allowed to read the token.

Under the above assumptions, it is possible to consider the setting in which $a = 2$, the user secret consists of m objects and the number of correct answers is $m/2$. As argued in the previous section, in this case the adversary cannot use

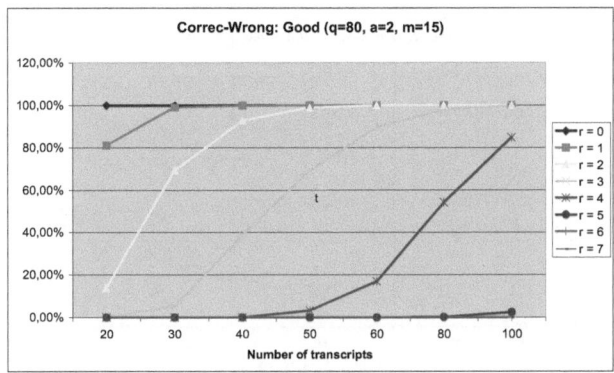

Fig. 6. Strategy Correct-Wrong (m=15, q=80, a=2). The adversary is provided with a number of transcripts in the range $\{20, 30, 40, 50, 60, 70, 80, 90, 100\}$. Percentage of good secrets extracted.

the attack technique described so far to extract good or valid transcripts. On the other hand, we can use the same technique presented in [2] to extract the user secret.

Although we focus on the Correct-wrong strategy, we show that the attack can be used also for the other authentication strategies presented in the paper.

Let us denote by $\alpha(k)$ the challenge for the the k-th transcript and let $\beta^{(k)}$ be the corresponding response. Since $a = 2$, $\alpha^{(k)}$ is a matrix consisting of 2 rows and $p = q/2$ columns. Let $(i_1^{(k)}, \ldots, i_p^{(k)})$ (resp., $(i_{p+1}^{(k)}, \ldots, i_q^{(k)})$) be the first (resp. the second) row of $\alpha^{(k)}$. In order to simplify the notation, we will omit the transcript number k when it is clear from the context.

We assign m different boolean variable $x_{i,1}, \ldots, x_{i,m}$ to each object o_i, with $i = 1, \ldots, q$. Intuitively, $x_{i,j} = 1$ implies that the j-th component of the user secret is o_i. Since each o_j appears in α exactly once, for every i, the i-th component of the user secret belongs either to the first or to the second row of α. For every $t = 1, \ldots, m$, i.e., for every component of the user secret, we define $\phi_{0,t} = x_{i_1,t} \vee \ldots \vee x_{i_p,t}$ and $\phi_{1,t} = x_{i_{p+1},t} \vee \ldots \vee x_{i_q,t}$

The adversary does know the user response $\beta = (\beta_1, \ldots, \beta_m)$, but she does not know which component of the response is correct and which one is wrong. On the other hand, the adversary knows that exactly $m/2$ out of m answers are correct. Let us define $A_m = \{\bar{a} = (a_1, \ldots, a_m) \in \{0, 1\}^m | w(\bar{a}) = m/2\}$, where $w(\cdot)$ denotes the Hamming weight of $\bar{a}$. Intuitively, if $a_i = 0$, the i-th answer contained in β is correct, otherwise is wrong.

Given the above notation, we can state that the following formula is satisfiable:

$$\psi = \bigvee_{(a_1,\ldots,a_m)\in A_m} \bigwedge_{j=1}^{m} (\phi_{\beta_j \oplus a_j, j} \wedge \neg\phi_{(1-\beta_j)\oplus a_j, j}). \tag{1}$$

Intuitively, the satisfiability of the above formula follows from the observation that: For a generic transcript (α, β) there exists a boolean array $(a_1, \ldots, a_m)$ that

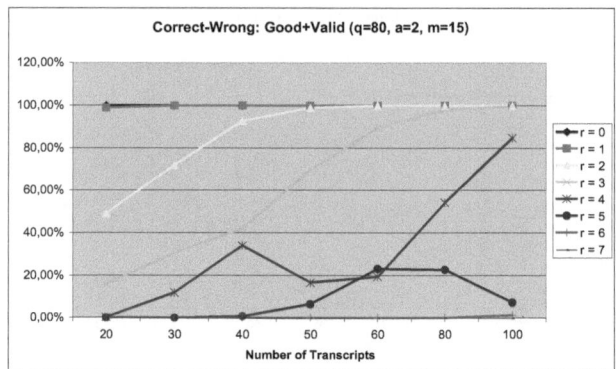

Fig. 7. Strategy Correct-Wrong (m=15, q=80, a=2). The adversary is provided with a number of transcripts in the range $\{20, 30, 40, 50, 60, 70, 80, 90, 100\}$. Percentage of good and valid secrets extracted.

identifies the correct and wrong answers. If the j-th answer in β is correct, i.e., $a_j = 0$, then the j-th component in the user secret belongs to the row identified by β_j (and, obviously, does not belong the the row identified by $1 - \beta_j$). Similar arguments apply for $a_j = 1$.

If the adversary is provided with t transcripts, the above formula has to be satisfied for each transcript. For this reason, if we denote by $\psi^{(k)}$ the Formula (1), properly rewritten for the k-th transcript, the following formula is satisfiable: $\psi' = \bigwedge_{k=1}^{t} \psi^{(k)}$.

Notice that the number of variables $x_{i,j}$ does *not* depend on the number of transcripts, i.e, all the formulas $\psi^{(k)}$ are written using the same variables.

The last constraint we need to consider is the fact that, each component of the secret consists of exactly one object. The above statement can be expressed by the following:

$$\epsilon = \bigwedge_{j=1}^{m} \bigvee_{i=1}^{q} (\neg x_{1,j} \wedge \ldots \wedge \neg x_{i-1,j} \wedge x_{i,j} \wedge \neg x_{i+1,j} \wedge \ldots \wedge \neg x_{q,j})$$

For any possible sequence of successful transcripts $((\alpha_1, \beta_1), \ldots, (\alpha_t, \beta_t))$ and for any possible secret σ, the formula $\mu = \epsilon \wedge \psi'$ is satisfiable. Notice that a truth assignment for μ might not represent the actual user secret. As an example, consider the case in which the adversary only holds a single transcript. Clearly the formula μ is satisfiable also in this case but there might exists multiple truth assignments. Clearly as the number of transcripts held by the adversary increases, the number of possible truth assignment for μ converges to 1, i.e., the actual user secret.

The attack just described can be easily modified for the Naive and the Correct-Random authentication strategies. In the former case, since all answers are correct, it is enough to consider the set $A_m = \{(0, 0, \ldots, 0)\}$. In the latter case, if r

is the number of random answers allowed by the scheme, the set A_m should be defined as: $A_m = \{\bar{a} = (a_1, \ldots, a_m) \in \{0,1\}^m | w(\bar{a}) \leq r\}$.

Currently we are implementing a test environment to experimentally evaluate the resiliency of the scheme presented w.r.t. this attack.

6 Conclusion and Future Work

In this paper we have presented a simple graphical PIN authentication mechanism that is resilient against shoulder surfing attacks. Our scheme is independent on the specific set of objects used to construct the challenges. The scheme may be implemented on low cost devices, does not require any special training for the users and requires a single round of interaction between the user and the terminal. We have argued that a secret consisting of 15 objects, e.g. letters, is enough to prevent the adversary to successfully authenticate even if the manages to obtain 36 transcripts.

The presented scheme can be also used for low-cost device authentication, e.g., RFID tag-to-reader or reader-to-tag authentication.

A number of extensions are possible for our scheme. An interesting variations is to authenticate the user if she answers correctly to a *specific* set of answers. Furthermore, is it possible to design a scheme in which the adversary manages to extract the user's secret *only if* she obtains a sequence of *consecutive* authentications? In the presented scheme, the adversary simply needs to obtain *any* sufficiently long sequence of authentications. If it should be possible to bind the secret extraction to the consecutiveness of the collected transcripts, in the real world the adversary may have very few chances of being successful.

Finally, we are currently experimentally evaluating the resilience of our scheme w.r.t. the SAT-based attack.

Acknowledgements

The authors thank Gene Itkis and Pino Persiano for their useful comments and suggestions.

References

1. Weinshall, D.: Cognitive authentication schemes safe against spyware (short paper). In: IEEE Symposium on Security and Privacy, pp. 295–300. IEEE Computer Society, Los Alamitos (2006)
2. Golle, P., Wagner, D.: Cryptanalysis of a cognitive authentication scheme (extended abstract). In: IEEE Symposium on Security and Privacy, pp. 66–70. IEEE Computer Society, Los Alamitos (2007)
3. Anderson, R.J.: Why cryptosystems fail. Commun. ACM 37, 32–40 (1994)
4. Steiner, J.G., Neuman, B.C., Schiller, J.I.: Kerberos: An authentication service for open network systems. In: USENIX Winter, pp. 191–202 (1988)

5. Haller, N.M.: The S/KEY one-time password system. In: Proceedings of the Symposium on Network and Distributed System Security, pp. 151–157 (1994)
6. McDonald, D.L., Atkinson, R.J., Metz, C.: One time passwords in everything (OPIE): Experiences with building and using stronger authentication. In: Fifth USENIX UNIX Security Symposium, Salt Lake City, Utah(USA) (1995)
7. Juels, A., Weis, S.A.: Authenticating Pervasive Devices with Human Protocols. In: Shoup, V. (ed.) CRYPTO 2005. LNCS, vol. 3621, pp. 293–308. Springer, Heidelberg (2005)
8. Juels, A.: Minimalist cryptography for low-cost rfid tags. In: Blundo, C., Cimato, S. (eds.) SCN 2004. LNCS, vol. 3352, pp. 149–164. Springer, Heidelberg (2005)
9. Lamport, L.: Password authentification with insecure communication. Commun. ACM 24, 770–772 (1981)
10. Matsumoto, T., Imai, H.: Human Identification through Insecure Channel. In: Davies, D.W. (ed.) EUROCRYPT 1991. LNCS, vol. 547, pp. 409–421. Springer, Heidelberg (1991)
11. Wang, C.H., Hwang, T., Tsai, J.J.: On the Matsumoto and Imai's Human Identification Scheme. In: Guillou, L.C., Quisquater, J.-J. (eds.) EUROCRYPT 1995. LNCS, vol. 921, pp. 382–392. Springer, Heidelberg (1995)
12. Matsumoto, T.: Human-computer cryptography: An attempt. In: ACM Conference on Computer and Communications Security, pp. 68–75 (1996)
13. Hopper, N.J., Blum, M.: A Secure Human-Computer Authentication Scheme. In: Carnagie Mellon University Technical Report. Vol. CMU-CS-00-139 (2000)
14. Hopper, N.J., Blum, M.: Secure Human Identification Protocols. In: Boyd, C. (ed.) ASIACRYPT 2001. LNCS, vol. 2248, pp. 52–66. Springer, Heidelberg (2001)
15. Katz, J., Shin, J.S.: Parallel and Concurrent Security of the HB and HB⁺ Protocols. In: Vaudenay, S. (ed.) EUROCRYPT 2006. LNCS, vol. 4004, pp. 73–87. Springer, Heidelberg (2006)
16. Grady, C.L., Mcintosh, A.R., Rajah, M.N., Craik, F.I.M.: Neural correlates of the episodic encoding of pictures and words. Proc. Natl. Acad. Sci. USA 95, 2703–2708 (1998)
17. Blonder, G.E.: Graphical passwords. Lucent Technologies Inc, Murray Hill, NJ (US), US Patent no. 5559961 (1996)
18. Perrig, A., Song, D.: Hash visualization: A new technique to improve real-world security. In: Proceedings of the 1999 International Workshop on Cryptographic Techniques and E-Commerce (1999)
19. Dhamija, R., Perring, A.: Déjà vu: A user study using images for authentication. In: IX USENIX UNIX Security Symposium, Denver, Colorado (2000)
20. Jensen, W., Gavrila, S., Korolev, V., Ayers, R., Swanstrom, R.: Picture password: a visual login technique for mobile devices. In: National Institute of Standards and Technologies Interagency Report, vol. NISTIR 7030 (2003)
21. Jensen, W.: Authenticating users on handheld devices. In: Proceedings of Canadian Information Technology Security Symposium (2003)
22. Real User Coorp.: Pass faces (1998), http://www.realuser.com
23. Jermyn, I., Mayer, A., Monrose, F., Reiter, M.K., Rubin, A.D.: The design and analysis of graphical passwords. In: Proceedings of the 8th USENIX security Symposium, Washington DC (1999)
24. Sobrado, L., Birget, J.C.: Graphical password. The Rutgers Scholar, an electronic Bulletin for undergraduate research 4 (2002)
25. Wiedenbeck, S., Waters, J., Sobrado, L., Birget, J.C.: Design and evaluation of a shoulder-surfing resistant graphical password scheme. In: Proceedings of Advanced Visual Interfaces AVI 2006, Venice, ACM Press, New York, NY (2006)

26. Roth, V., Richter, K., Freidinger, R.: A pin-entry method resilient against shoulder surfing. In: CCS 2004: Proceedings of the 11th ACM conference on Computer and communications security, pp. 236–245. ACM Press, New York (2004)
27. University of British Columbia (Ubcsat, the stochastic local search sat solver), http://www.satlib.org/ubcsat
28. Kumar, M., Garfinkel, T., Boneh, D., Winograd, T.: Reducing shoulder-surfing by using gaze-based password entry. In: Symposium On Usable Privacy and Security (SOUPS) (2007)
29. Suo, X., Zhu, Y., Owen, G.S.: Graphical passwords: a survey. In: Proceedings of 21st Annual Computer Security Application Conference (ACSAC 2005), December 5-9, 2005, Tucson AZ (US), pp. 463–472 (2005)
30. Graphical Password Project: Falces (1998), http://www.realuser.com

Software Cannot Protect Software: An Argument for Dedicated Hardware in Security and a Categorization of the Trustworthiness of Information

Matthew Judge, Paul Williams, Yong Kim, and Barry Mullins

Air Force Institute of Technology
2950 Hobson Way
Wright Patterson AFB OH 45433, USA
{matthew.judge,paul.williams,yong.kim,barry.mullins}@afit.edu

Abstract. There are many current classifications and taxonomies relating to computer security. One missing classification is the *Trustworthiness of Information* being received by the security system, which we define. This new classification along with *Timeliness of Detection* and *Security level of the Security System* present motivation for hardware-based security solutions. Including hardware is not an automatic solution to the limitations of software solutions. Advantages are only gained from hardware through design that ensures at least *First-hand Information*, dedicated monitors, explicit hardware communication, dedicated storage, and dedicated security processors.

1 Introduction

As security takes on ever increasing importance in today's connected, digital world; security solutions incorporate new, dedicated hardware at an increasing rate [1,2,3,4,5,6,7,8,9,10,11,12,13]. Though these works and many others investigate the incorporation of hardware into designs to gain different advantages, little work has been dedicated to understanding what precisely can be accomplished with hardware that cannot be accomplished solely with software solutions. Though many people believe a hardware-based solution is necessary to achieve effective security, little or no work exists demonstrating that this is true. The first and most obvious question to be asked is whether hardware solves the shortcomings and vulnerabilities of software based solutions. Exploring this question leads to a critical answer: *Not necessarily*. This work then, attempts to capture the necessary design elements for creating hardware that overcomes the weaknesses of purely software-based solutions. To aid in defining these requirements, we propose a classification for the *Trustworthiness of Information* and show that the necessary level of trust, *First-hand Information*, can *only* be achieved by properly designed hardware. Complete security systems will integrate these key hardware components with security software as needed.

J.A. Onieva et al. (Eds.): WISTP 2008, LNCS 5019, pp. 36–48, 2008.

2   Current Security Classifications

Significant work has been published on categorizations, classifications, and taxonomies for computer security. Bazaz and Arthur present a taxonomy of vulnerabilities [14]. Axelsson develops a taxonomy of detection methods [15], that Williams extends [16]. Kuperman classifies both the goals of detection and the timeliness of detection [17], and Stakhanova et al. work towards a taxonomy of intrusion detection system responses [18]. Mott presents work into classifying the level of security that the *security system* maintains for itself [5]. All of these different classifications provide valuable insight for working in the security field. One critical classification missing from these is the *Trustworthiness of Information*, which we develop in Sec. 4.3. Mott's work and Kuperman's timeliness of detection classification are both integral to the discussion of why hardware is necessary both on their own and how they relate to and are influenced by the *Trustworthiness of Information*. We discuss each in greater depth here.

Kuperman's notation categorizes time into an ordered sequence of events [17]. He defines the set of all events that can occur in the system, E, the subset of all malicious events, B, $B \subseteq E$, and three events a, b, c such that $a, b, c \in E$ and $b \in B$. Given the notation t_x to represent the time of event x occurring and $x \to y$ representing a causal dependence of y upon x we assume the three events are related such that $a \to b \to c$ yielding the relationship, $t_a < t_b < t_c$ must be true. Note that although $x \to y$ represents a causal dependence it does not necessarily mean that x is the direct cause of y. Kuperman uses $D(x)$ to represent the detection of an event x.

With this notation defined, Kuperman presents four main timeliness categorizations: real-time detection, near real-time detection, periodic detection, and retrospective detection. We discuss the first two here, which represent detection categories we hope to improve through our research.

Real-time Detection. The detection of a bad event, b, occurs while the system is operating and before events dependent on b occur, requiring the order

$$t_b < t_{D(b)} < t_c \tag{1}$$

Near Real-time Detection. The detection of a bad event, b, occurs within some predefined time step δ, either before or after t_b.

$$|t_b - t_{D(b)}| \leq \delta \tag{2}$$

Kuperman comments that this timeliness categorization should be independent of the underlying hardware and the rate of event occurrence. Although this goal is desirable for a software-based solution, it relies on assumptions of trustworthiness and lack of vulnerabilities in this underlying hardware. With today's computer hardware this independence is unobtainable. Rutkowska's attack, discussed in Sec. 3, provides a specific example of why hardware cannot be blindly trusted. If hardware cannot automatically be trusted it *must* be considered in security measurements.

An often overlooked aspect of a computer security monitor is the security of the monitor itself. This security is a critical aspect of a security system, since compromising the monitors can effective render the security system useless. Mott presents a classification of the security of the monitors creating eight levels of monitoring system security [5] presented here.

Open. This worst case scenario occurs when the monitored system has knowledge of the monitor and shares information with the monitor without any security mechanisms present.

Soft Security. This level of monitor security is equivalent to *open* with software used to secure the monitor. Both of these levels tend to contain monitors on a uniprocessor host-based intrusion detection system.

Passive Security. The monitor operates without the monitored system necessarily knowing it is there. To compromise such a system, information about how the monitor analyzes gathered state data must be known. Prime examples of this level of security include most network Intrusion Detection Systems (IDSs) where only network traffic is monitored. Specific information passed over the network has the potential to disable the system, but there are no direct avenues of attack.

Self Security. Similar to both *open* and *soft* security systems, the monitored system shares information with the monitor. The manner in which the monitor operates provides it with security, requiring the monitored system to be compromised before the monitor can be compromised. An example of this level of security is Williams' CuPIDS [16].

Loose-hard Security. The monitored system again has knowledge and coordinates with the monitor, sharing information, but dedicated hardware mechanisms protect key portions of the security system from compromise such as with hardware-based return address stacks [19].

Semi-hard Security. The monitored system's knowledge of the monitor is extremely limited. To provide this level of security the monitor cannot execute on the same processor core as the monitored software and communications happens through mechanisms like unmaskable interrupts that are kept to a minimum. Compromise can only occur via code controlling synchronization signals to the monitor, which would cause the monitor to operate in a diminished capacity.

Strict-hard Security. This security level adds to the requirements of *semi-hard* security by requiring only hardware connections to the monitor and removing synchronization signals to the monitor. The monitor must be able to gather its own state information to remove dependence of the monitor on the monitored system. Two examples of this level of security are CoPilot [6] and Independent Auditors [4].

Complete Security. This level of security is the ideal secure case, used as a theoretical comparison point. In reality, such a monitoring system would have no contact with the production system, negating it's usefulness.

Mott notes that with many of these levels of security, there is a tradeoff between the security of the monitor and the ease with which state information can be gathered from the monitored system [5]. One critical piece of information overlooked by these categories is the trustworthiness of the information that the monitor is receiving. Although technically the monitor itself is not corrupted, the effects can be equivalent. For example, a Supervisory Control And Data Acquisition (SCADA) System controlling critical infrastructure such as the electrical grid, could be manipulated to perform undesirable actions, without ever compromising the SCADA System. This can still be accomplished by an attacker who can only manipulate the information being received by the SCADA System. For instance, if an attacker can manipulate the information feeding the SCADA System, telling it that there is a massive overdraw on the electrical grid, they can affect SCADA System responses such as causing a rolling blackout. This is accomplished without specifically corrupting the SCADA system to do so. The SCADA System would respond correctly to the environment it believes exists, not the actual environment. A simpler exploit corrupting the information being passed to monitors is a denial of service (DoS) attack. If the SCADA system does not receive readings from sensors monitoring critical sections of the system, it will be unable to respond to parameters out of acceptable ranges. This could quickly compound into catastrophic failure.

Although this issue is acknowledged in a number of works [6,16,17], we have not found research that investigates this aspect. Our research explores this aspect of the monitor's security. Rutkowska presents methods for corrupting the memory access of the PCI Bus without affecting the processor's access to memory [20] which is discussed in more detail in Sec. 3. This exploit highlights the importance of this aspect of classification for the security of the monitoring system. CoPilot [6], one of the examples Mott identifies as being *strict-hard* security, is defeated by this attack because of its security weakness on this new axis of categorization. We present an independent axis for categorizing the security of the monitor relating to the trustworthiness of the monitored data: how far removed the monitor is from what it is monitoring. This classification is defined in Sec. 4.3.

3   Defeating Hardware-Based RAM Acquisition

As the previous section began to develop, the ability to falsify the information a security monitor receives corrupts the integrity of the security system. One prominent example of this exploitation is Rutkowska's defeat of hardware based random access memory (RAM).

Rutkowska discusses both software and hardware approaches to memory acquisition with the claim that the hardware based approaches are superior to that of software based solutions [20]. She cites non-persistent malware as motivation for needing memory acquisition and she presents a number of known exploits of software memory acquisition by code running at the same privilege level as the acquisition software. One specific example of such an exploit is the FU Rootkit [21]. Rutkowska notes that these software memory acquisition tools require additional software on the target machine, which she claims violates the

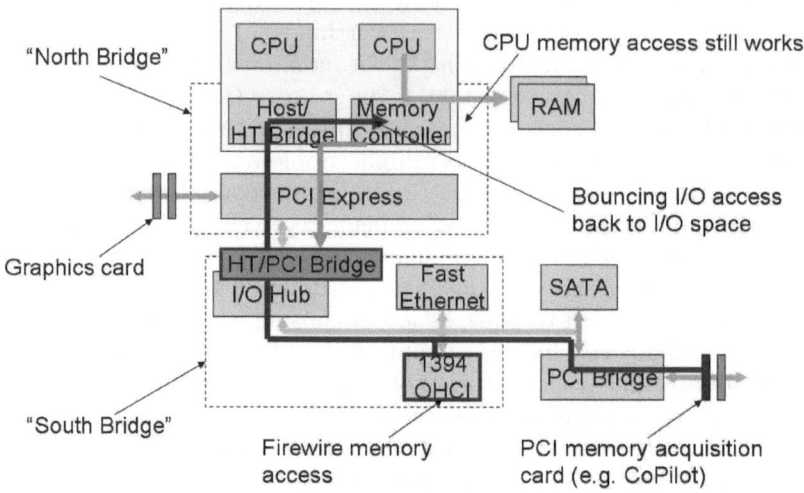

Fig. 1. Rutkowska's Defeat of Hardware Based RAM Acquisition [20]

forensic tool requirement not to write data to the targeted machine. She then extols the virtues of hardware based solutions, setting her readers up for her defeat of this "superior" memory acquisition method.

Rutkowska delivers three levels of compromise to hardware based memory acquisition devices such as CoPilot [6] and Tribble [1]; each building upon the same basic exploit with increasing levels of damage. This exploit, depicted in Fig. 1 involves configuring the north bridge on a system to map arbitrary ranges of physical memory to I/O space. This remapping denies memory access to peripheral devices for the specified physical memory range while not affecting the memory access of the processor(s). This allows an exploit to execute correctly, while hiding the exploit's presence from the hardware-based acquisition tool. These levels range from a denial of service to an attack that provides the monitor with false data, completely masking the compromise from the monitor.

The exploits Rutkowska presents show definitively that current hardware based memory acquisition devices, such as those that plug in to a firewire port or as a PCI device, are not reliable. The lesson to be taken from her work is not that hardware cannot do a better job of providing security features, rather that hardware is not a magic bullet; it does not automatically improve security. This work highlights that many current hardware solution are missing an important aspect of the capability and security of the monitoring system. This provides substantial motivation to explore the trustworthiness of the information being received by a security monitor. This critical axis of security for a monitor, though acknowledged in numerous works [1,6,16,17] is not well understood and not clearly defined. Section 4.3 provides definitive categorization along this axis to aid future work in security related fields understand what is required to provide truly reliable security monitoring.

4   Why Hardware?

Most current security systems for computers are based largely on software systems. Numerous flaws and vulnerabilities have been exposed and even exploited in these different software solutions. Compromise of protected code via rootkits [22] represents one of the most prevalent exploits. Recent work has begun exploring different hardware based approaches to security [5,6,12] with many people coming to believe that we cannot solely use software to protect software and only hardware, coupled with software, can do that job successfully [1,2,3,4,5,6,7,8,9,10,11,12,13]. Though a number of advantages to hardware over software have been suggested, we found no research discussing what precisely makes hardware a significant improvement over software and just what capabilities hardware provides that software cannot. We present here a number of key advantages achievable through the use of hardware.

Reduced Avenues of Attack. Separate monitoring hardware can strengthen the security of the monitor by reducing the extent of the coupling between the security and production systems.

Trustworthiness of Information. Correctly designed hardware guarantees that the monitor receives valid data from the production system, something that we show software incapable of doing.

Additional/Different Information Available. Mott's research explores a number of pieces of information that can be gathered through hardware primitives and leveraged to increase the overall security of the system [5]. These hardware primitives include information such as the program counter, instruction traces, and added visibility into memory.

Timeliness of Detection. The ability to guarantee real-time detection, as defined in Sec. 2, requires the ability to guarantee that the monitor will execute with the ordering of (1). Dedicated monitors are necessary to accomplish this.

In the rest of this section we develop justification for needing capabilities beyond what software can provide and explore each of these advantages in greater detail. We define what is required of hardware to overcome the vulnerabilities of software and provide significantly improved performance.

4.1   Vulnerabilities of Software Security Systems

There are a number of vulnerabilities inherent in software-based security. Two critical vulnerabilities are the inability to guarantee real-time monitoring in standard commercial operating systems, even on a multiprocessor system, and the inability to protect the integrity of the security system once the production system has been compromised. The first vulnerability is evidenced by the fact that scheduling of processes on both uniprocessor and multiprocessor systems does not make any guarantees on precise ordering or timing of when a specific process gets time on a processor. Work such as CuPIDS changes this standard paradigm to guarantee monitored processes run in lock step with the monitoring process [16] and overcome this first critical vulnerability of software security systems. Despite CuPIDS' ability to overcome this vulnerability, it cannot protect itself once the kernel has been compromised.

The specific point where software loses the ability to protect other software is when faced with exploitation of a vulnerability in privileged code. Once an attack can gain access through such a vulnerability, they have access to any piece of software in the system and can modify both data and executable code. This allows for changes in both user applications and the operating system itself, compromising the security of the security system itself. This can be accomplished through modification to the security software itself or by modifying the operating system to interact with the security software in another manner, such as reducing its privilege level. Note that exploitation of vulnerabilities in privileged code provides two main avenues of attack into the system. The more obvious method of attacking the security software itself is to degrade or interrupt its capabilities described above. The other avenue of attack is corrupting the information that is being sent to the security software.

This second issue is the general method that rootkits use to remain undetected. They interpose themselves between processes by taking control when there is a library function or system call. By controlling what information is passed back to the monitoring process the rootkit can neutralize the security software without directly modifying it [22].

4.2   Advantages of Hardware

The vulnerabilities of software discussed above show clear need for a security solution that can overcome these vulnerabilities. Does hardware provide protection from these attacks? Not necessarily. Hardware can provide increases in protection, but only if appropriately designed into the system's architecture. Two key factors in designing hardware that can enhance these areas of security are where we connect the security hardware to the system and how we make those connections. Where we connect controls the *Trustworthiness of Information* as well as influences the *Timeliness of Detection*. The next two subsections explore these advantages in greater detail. How the security hardware is connected impacts the amount of information available to the security system and defines the only avenues of direct attacks on the security system. By limiting the physical pathways between the production system and the security system to specific hardware primitives, the attack surface is significantly reduced. These primitives can also provide access to key information which is unobtainable via software-based solutions. Both aspects of hardware primitives are discussed in Sec. 4.5.

4.3   Trustworthiness of Information

Although the need for the monitor to receive accurate data is understood, there is no real framework for understanding what precisely is needed to accomplish this. Towards this end we define a new axis categorizing the trustworthiness of the information being received by the monitor. This axis of trustworthiness stands as its

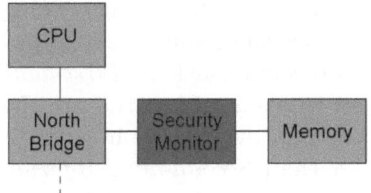
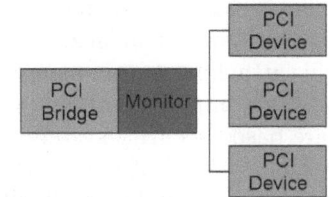

Fig. 2. Immediate Information: Security Monitor placed inline between main memory and the memory controller

Fig. 3. First-hand Information: Security Monitor placed on a shared bus, vulnerable to Denial of Service from excessive device traffic

own contribution and should be considered when attempting to provide an accurate, secure monitoring device of any sort. By creating this categorization we set important bounds on what exactly affects the trustworthiness of the information.

Immediate Information. (Fig. 2) With immediate access to what is being monitored we can insure the monitor is receiving true data. This immediate categorization represents a specific form of first-hand information where the monitor is inline, directly between what is being monitored and its interaction with the system. While this level of trustworthiness is certainly the most definitive method for ensuring the monitor's security, it leads toward a design with individual monitors on every single hardware component, thus requiring a complete redesign of all aspects of a system.

First-hand Information. (Fig. 3) This level of trustworthiness represents a monitor that has direct access to the data being output from some device. Depending on the specific design of the architecture being monitored, this level of trustworthiness will likely be equivalent to *Immediate Information*. However, a shared bus architecture could be vulnerable to a denial of service (DoS) exploit. This would be accomplished in much the manner that someone would have trouble listening to another's conversation in a crowded room.

Second-hand Information. This level of trustworthiness encompasses any monitor that relies on some intermediary mechanism, such as hardware or software components, to pass it the data it is attempting to monitor. Although each additional mechanism relied upon reduces the trustworthiness into third-hand information and so forth with a continually lessening level of trustworthiness. For simplicity we group all levels of trustworthiness that cannot guarantee accurate monitoring into this category of second-hand information. Unless any and all mechanisms being relied upon to pass the monitor data can be guaranteed secure, this presents an avenue of attack for corrupting the monitor be feeding it false data. Figure 1, on page 40, shows a PCI-based memory acquisition tool, such as CoPilot [6], that must trust the PCI bridge, the south bridge, and the north bridge; trust which Rutkowska's research demonstrates as unwarranted [20].

It is this previously undefined axis of the monitor's security that is being exploited by Rutkowska's attack. Our research defines the requirement to protect against this attack: monitors must be capable of receiving at least *First-hand Information*. Two important things to note about this axis of security are that 1) all software based security systems on a uniprocessor system are inherently unable to achieve a level of trustworthiness better than *Second-hand Information* since they must rely on data controlled by the operating system and 2) even software based solutions designed to operate within a multiprocessor system, such as CuPIDS [16], must still rely on the trustworthiness of main memory and therefore receive no better than *Second-hand Information*. In order to ensure accurate monitoring, the monitor needs to have access to at least *First-hand Information* of the data being produced, any intermediate devices provide the possibility of the data being manipulated before reaching the monitor. Therefore at very least we need monitoring or interaction points at each of the bridges in the system, i.e. any device that passes information from one part of the system to another.

4.4   Timeliness of Detection

Another aspect of monitor placement is the speed with which a monitor can detect an attack. One of the areas where the speed of a device far exceeds the speed of the buses that pass information to and from it is the processor(s). To accomplish real-time monitoring as defined in Sec. 2, monitors will need to be closer to the main processor than system bridges will allow. One such example of this is a hypothetical purely cache based attack [6]. Such an attack will be able to do its damage before detection, since detection is only possible with access to a present view of cache. Even if we accept near real-time monitoring capabilities, Kuperman's δ value in (2) will be significantly smaller for a monitor that is located on-chip.

4.5   Hardware Primitives

The manner in which we connect monitors to the system plays a significant role in enhancing both the security of the production system and the security of the security system. By limiting connections between the monitor and production systems and remaining within Mott's *Semi-hard* security level the only avenues of directly attacking the security system are the hardware primitives that bridge the monitors and production system. As long as no primitives allow for modification of the monitoring system's code, we maintain a greatly reduced attack footprint for the security monitor. At the same time, these hardware primitives can offer direct access to information previously difficult to obtain and even provide access to information not accessible through any software methods. Mott presents a number of hardware primitives that can be leveraged in [5]. The two main areas of interest for creating hardware security (and security in general) have been attempts to monitor processes running on the production system, mainly through various memory introspection techniques [1,2,3], and monitoring the incoming network traffic as it enters the system [8,9,10,11,12,13].

4.6   What Do We Mean by Hardware Security?

To this point we have left the definition of hardware security somewhat up in the air. All computer systems contain a mix of hardware and software and only a limited amount is accomplished with purely hardware. To create a security system purely in hardware would significantly hamper the flexibility and modifiability of such a system reducing the number of future attacks to which a system could potentially respond. Solutions such as a field programmable gate array (FPGA) can be used to extend software flexibility into hardware, though it does require performance tradeoffs and is not pivotal to this aspect of our discussion. However, a pure hardware solution is not our goal when we talk about hardware-based security. The key component of hardware-based security is the *communication between the production system and the security system*. Whether a specific monitor is pure hardware, a FPGA, or software running on some combination of hardware that remains separate from the production system hardware, what qualifies a security component as hardware-based is that connection back to the production system. Note that an important result of this definition is that a hardware-based security solution requires physically separated memory. This is not to say that pure hardware or at least FPGA solutions will not be required in some instances to provide fast enough response. Areas where high-speed detection is crucial will almost certainly benefit from pure hardware solutions. One predominant example of this is the network IDS field where research has shown benefits from hardware solutions [10,11,13,23].

4.7   Hardware/Software Interaction

With the key component of using hardware being the communication between the production system and the security system, software can be employed on a separate security processor. This allows a full-fledged software security operating system to run on such a dedicated security processor. Mott et al. explore this interaction, pointing to the hardware monitors as decoupling production and security software [24]. This software can perform management and communication roles between elements of the security system so long as there is no access to modify the software via the production system. With the inclusion of dedicated security system I/O, via some variant of a communications port, the software can be modified and updated as needed to respond to future threats.

5   Specific Requirements for Achieving Benefits from Hardware

So far we have discussed the different advantages of dedicated hardware for security solutions and discussed what is required to achieve these advantages. Here we explicitly define these requirements for dedicated hardware. By designing to these requirements, it is possible to design a comprehensive security solution that achieves the advantages of hardware previously explored. These requirements are:

First-hand Information. of all monitored information: This level of trusted information guarantees accurate monitoring of what is happening in the system. Without this level of trusted information security solutions are vulnerable to being denied access to the information or even fed false information. This vulnerability provides a route to compromise the effectiveness of the security system, without the need to compromise the security system itself.

Dedicated Monitors. for parallel, concurrent monitoring: To protect against potential timing attacks monitors must be able to run concurrently with what they are monitoring to allow the possibility guaranteeing of Kuperman's *real-time detection* [17]. Any monitor which does not run concurrently with its target must ensure that it runs often enough to be impervious to timing attacks. In a software-based solution this becomes infeasible due to the performance penalty of continuous context switching. Dedicated hardware monitors remove the burden on production resources and keep performance degradation to a minimum [7].

Explicit Hardware Communication. between the production and security systems: By limiting communication between the production and security systems to hardware pathways, we reduce avenues of attack upon the security system to these explicitly defined pathways. Without modifiable communication pathways, the ability to corrupt these pathways is reduced. These limited pathways provide a clear set of attack avenues which can be understood and protected.

Dedicated Storage. of security code and data: Without dedicated, separate security storage we leave software communication pathways present in the system. These communication pathways represent a significant avenue of attack to be exploited. Any software-based separation becomes vulnerable to a root-level compromise of the production system. Separate storage which cannot be directly modified by the production system provides a more reliable method of protecting the security code and data.

Dedicated Security Processor. for controlling and coordinating the security mechanisms: Though not explicitly a requirement for gaining security capabilities, a dedicated security processor is included here for the coordination and communication abilities it can provide. This separate processor will allow for a secured security control center when coupled with these other requirements. It will provide the ability to modularly add security mechanisms into a security backplane. An important aspect of this ease of modularity is the ability to combine both network IDSs and host-based IDSs into a combined, complete IDS which can leverage combined knowledge from each to provide more flexible and effective response.

6   Conclusion

The use of hardware is necessary to provide quality security solutions. Short of verifying the trustworthiness and security of all software and hardware mechanisms in the chain from the monitored information back to the monitor, *First-hand Information*, that requires dedicated hardware to achieve, is the only way to guarantee the security monitor is not fed false data. As computer security systems become more reliant on dedicated hardware, the need for a clear understanding of the necessary design requirements to overcome inherent software security vulnerabilities is essential. This work provides a basis for this understanding by defining the advantages that can be gained from hardware, and the necessary design to achieve them.

References

1. Carrier, B.D., Grand, J.: A hardware-based memory acquisition procedure for digital investigations. Digital Investigation 1 (2004)
2. Özdoganoglu, H., Vijaykumar, T.N., Brodley, C.E., Kuperman, B.A., Jalote, A.: Smashguard: A hardware solution to prevent security attacks on the function return address. IEEE Transactions on Computers 55 (2006)
3. Gordon-Ross, A., Vahid, F.: Frequent loop detection using efficient non-intrusive on-chip hardware. In: CASES 2003: Proceedings of the 2003 international conference on Compilers, architecture and synthesis for embedded systems (2003)
4. Molina, J., Arbaugh, W.: Using independent auditors as instrusion detection systems. In: Information and Communications Security: 4th International Conference (December 2003)
5. Mott, S.: Exploring hardware-based primitives to enhance parallel security monitoring in a novel computing architecture. Master's thesis, Air Force Institute of Technology (March 2007)
6. Petroni, N.L., Fraser, T., Molina, J., Arbaugh, W.A.: Copilot-a coprocessor-based kernel runtime integrity monitor. In: Proceedings of the 13th USENIX Security Symposium, pp. 179–194 (2004)
7. Williams, P.D., Spafford, E.H.: Cupids: An exploration of highly focused, coprocessor-based information system protection. Computer Networks, 51 (April 2007)
8. Song, H., Lockwood, J.W.: Efficient packet classification for network intrusion detection using FPGA. In: FPGA 2005: Proceedings of the 2005 ACM/SIGDA 13th international symposium on Field-programmable gate arrays (2005)
9. Yi, S., koo Kim, B., Oh, J., Jang, J., Kesidis, G., Das, C.R.: Memory-efficient content filtering hardware for high-speed intrusion detection systems. In: SAC 2007: Proceedings of the 2007 ACM symposium on Applied computing (2007)
10. Gonzalez, J.M., Paxson, V., Weaver, N.: Shunting: a hardware/software architecture for flexible, high-performance network intrusion prevention. In: CCS 2007: Proceedings of the 14th ACM conference on Computer and communications security (2007)
11. Hutchings, B.L., Franklin, R., Carver, D.: Assisting network intrusion detection with reconfigurable hardware. In: FCCM 2002: 10th Annual IEEE Symposium on Field-Programmable Custom Computing Machines, vol. 00 (2002)
12. Hart, S.: APHID: Anomoly processor in hardware for intrusion detection. Master's thesis, Air Force Institute of Technology (March 2007)
13. Bu, L., Chandy, J.A.: FPGA based network intrusion detection using content addressable memories. In: Proceedings - 12th Annual IEEE Symposium on Field-Programmable Custom Computing Machines, FCCM 2004, CA 90720-1314, United States, April 2004, IEEE Computer Society, Los Alamitos (2004)

14. Bazaz, A., Arthur, J.D.: Towards a taxonomy of vulnerabilities. In: Hawaii International Conference on System Sciences (2007)
15. Axelsson, S.: Intrusion detection systems: A survey and taxonomy. Technical report, Chalmers University of Technology (March 2000)
16. Williams, P.D.: CuPIDS: Increasing Information System Security Through The Use of Dedicated Co-Processing. PhD thesis, Purdue University (August 2005)
17. Kuperman, B.A.: A Categorization of Computer Security Monitoring Systems and the Impact on the Design of Audit Sources. PhD thesis, Purdue University (2004)
18. Stakhanova, N., Basu, S., Wong, J.: A taxonomy of intrusion response systems. Technical Report 06-05, Department of Computer Science, Iowa State University (2006)
19. Lee, R.B., Karig, D.K., McGreggor, J.P., Shi, Z.: Enlisting hardware architecture to thwart malicious code injection. In: Hutter, D., Müller, G., Stephan, W., Ullmann, M. (eds.) Security in Pervasive Computing. LNCS, vol. 2802, pp. 237–252. Springer, Heidelberg (2004)
20. Rutkowska, J.: Beyond the CPU: Defeating hardware based RAM acquisition (February 2007), http://invisiblethings.org/papers.html
21. Rootkit, F.U.: http://www.rootkit.com/project.php?id=12
22. Levine, J., Grizzard, J.O.H.: A methodology to detect and characterize kernel level rootkit exploits involving redirection of the system call table. In: Information Assurance Workshop, 2004. Proceedings. Second IEEE International, April 2004, pp. 107–125 (2004)
23. Tummala, A.K., Patel, P.: Distributed ids using reconfigurable hardware. In: 21st International Parallel and Distributed Processing Symposium, IPDPS 2007. Institute of Electrical and Electronics Engineers Computer Society, Piscataway, NJ 08855-1331, United States (March 2007)
24. Mott, S., Hart, S., Montminy, D., Williams, P., Baldwin, R.: A hardware-based architecture to support flexible real-time parallel intrusion detection. In: Proc. 2007 IEEE International Conference on System of Systems Engineering (2007)

Probabilistic Identification for Hard to Classify Protocol

Elie Bursztein[*]

LSV, ENS Cachan, CNRS, INRIA
`eb@lsv.ens-cachan.fr`

Abstract. With the growing use of protocols obfuscation techniques, protocol identification for Q.O.S enforcement, traffic prohibition, and intrusion detection has became a complex task. This paper address this issue with a probabilistic identification analysis that combines multiples advanced identification techniques and returns an ordered list of probable protocols. It combines a payload analysis with a classifier based on several discriminators, including packet entropy and size. We show with its implementation, that it overcomes the limitations of traditional port-based protocol identification when dealing with hard to classify protocol such as peer to peer protocols. We also details how it deals with tunneled session and covert channel.

Keywords: payload analysis, header discriminator, p2p, traffic classification.

1   Introduction

The use of protocol identification as a defense against unwanted traffic such as P2P (peer to peer) and Malware, has received a lot of attention lately. It might appear that there is a simple identification technique to classify traffic : assuming that protocols will use well-known ports such as the one assigned by the IANA [18]. This is, however, not reliable anymore. A recent study [25] reports that in a large university about 40% of the traffic failed to be classified by this heuristic. In particular, many P2P (Peer to Peer) protocols use obfuscation techniques to avoid detection [21]. They do not use static well-known port numbers, but rather dynamically use available port numbers. They also masquerade themselves by using ports reserved for other applications and encrypt their packet payloads [29]. For example, an Edonkey node can use the port 80, typically reserved for HTTP for its own communication, allowing it to confuse firewalls, packet shaper and passive network monitors [21]. Moreover prohibited software and botnets try to deceive network security devices by tunneling their connections into other protocol such as ICMP or HTTP [7].

In this paper, we shall show that our advanced protocol analysis allows to improve protocol identification and detect tunneled and covert session. In

---

[*] PhD student, supported by a DGA grant.

J.A. Onieva et al. (Eds.): WISTP 2008, LNCS 5019, pp. 49–63, 2008.

particular, we demonstrate through our passive network monitor prototype *NetAnalyzer* [6] evaluation, that our analysis if effective against two of the most populars P2P network: Edonkey and BitTorrent while remaining sufficiently efficient to be used online.

We choose to implement our technique in a passive network monitor because despite the great benefits provided by an advanced protocol analysis, every public passive network monitors we are aware of, including NTop [10] and Iptraf [19], still solely rely on a port-based heuristic for traffic classification. Additionally to traffic classification, our analysis reports valuable information for network assessment such as software products and protocol versions that can be used as contextual information for security evaluation.

The main contribution of this paper is a probabilistic identification analysis that combines a packet payload analysis, a packet classifier and the port heuristic. The payload analysis is based on signature. The classifier uses multiple packet discriminators: packet entropy, packet size, and packet intervals. To our knowledge, this is the first work that combines multiples identification techniques to return an ordered list of probable protocols for a given session. It seems to be also the first to add to signature and techniques a confidence value.

The rest of this paper is organized as follows. In Sect. 2, we will survey related work and in Sect. 3 we will give a straightforward example of how the analysis works. This example is used as a guideline for the rest of the paper. Sect. 4 presents our probabilistic identification analysis. Sect. 5 details how tunneled session are handheld. Sect. 6 shows how covert channel are detected. Sect. 7 details a specific application of the advanced analysis to passive monitoring: file masquerading detection. In Sect. 8 we evaluate the accuracy and the speed of the analysis against P2P traffic. Finally, we will conclude in Sect. 9.

We emphasize that even thought our discussion focus on passive network monitor, our analysis is not limited to this use. The ability to have a reliable protocol identification and session contextual information such as software version are also valuable for Firewall and NIDS (Network intrusion detection system) and traffic shaper.

2 Related Work

Ranking algorithms are a common for information retrieval [39,15]. For instance Google use the well known Google Page rank algorithm [30]. Beside payload and packets discriminators, others techniques exists. They are either unusable for online analysis [42], or specific to a type of protocol such a P2P detection [9,21,16]. Protocol identification through payload analysis is used for Q.O.S enforcement in CISCO router [3] and Linux Netfilter [35]. It is also used in Bro NIDS [11] to instantiate appropriate decoders. The use of automatic signature generator has received a lot of attention [22,37,17,41] because it removes the burden to create signature by hand. Tools such as Polygraph [28] and Hamsa [24] are used to create attack signature automatically. Attack against automatic signature generator have been studied in [8]. Some of *NetAnalyzer* signatures are taken from the Nmap [14], and L7 filter [35] databases. This technique uses packets headers to build a classifier [27] based on discriminators. Some discriminators work on packet payload for example packet entropy, [36], and character frequency [41]. Others, such as the packet size or time interval between packets [5] are payload independent and focus on packet headers. In [23], a combination of six discriminators is used to detect intrusion in HTTP CGI. Malwares and P2P clients, even Skype [4], use many techniques to avoid protocol identification. Packet padding [13] is used in Emule [34] to avoid traffic classifier. Popular BitTorrent clients use the "Message Stream Encryption" scheme [2] which is specifically designed to defeat protocol identification. Botnet use ICMP [7] and DNS [40] covert channel. Tunneling a protocol trough a proxy [12], is a popular technique to defeat protocol prohibition e.g IRC, Instant Messaging.

3   Identification Result Example

This example highlights the two most relevant benefits provided by our analysis namely: the accurate ranked protocol identification, and the session advanced information reporting.The example is a HTTP session reported by *NetAnalyzer* (figure 1). Each session report is composed of four parts: the summary (line 1), the traffic information (line 2), the advanced information (line 3 to 10) and the identification details (line 11 and 12);

```
(1)http (94%): x:1052 -> y.:2080 F:0/2 R::2/0 R:0/2 E:0/0 TCP CLOSED RST
(2)[Traffic] I:1 Kb/s (3pkt) O:37 Kb/s (20pkt) [Distance] C:local S:7
(3)[Protocol]:http (1.1) HyperText Transfer Protocol - RFC 2616
(4)[File] request:www.xxx.org/vip.html Ref:"http://xxx"
(5)[File] content: extension:.html familly:text   (X)HTML
(6)[File] request:www.xxx.org/hello.gif Ref:"http://xxx"
(7)[File] content: extension:.jpg familly:image
(8)[Server] Apache httpd 2.0.52
(9)[Client] browser Internet Explorer 6.0 Windows XP
(10)[Client] proxy squid 2.5.STABLE4-20031106
(11)[Guessed protocol] http:94% Port:0% Class:100% Patt:100%
(12)[Guessed protocol] autodesk:9% Port:100% Class:n/a Patt:0%
```

Fig. 1. *NetAnalyzer* Session output example

3.1   Protocol Identification

This first major benefit of our identification analysis, is the ability to identify the correct protocol regardless of its and to provide the accuracy probability of its identification.

The protocol identified with the highest probability is displayed on the leftmost part of the first line. Here it is the *http* protocol with a probability of 94%. It is not 100% because the server port 2080 is not the standard one: 80. Hence there is a conflict between the port heuristic result and the payload analysis result. The list of all possible protocols for the session, is presented at the end of the report (line 11 and 12). Each line gives the protocol name, its probability and details the result of each technique used. Intuitively the probability of 94% for the HTTP protocol results of the six signatures positive match along with the port heuristic's negative result and the classifier 100% positive result. The classifier score is 100% positive because every 23 packets of the session match HTTP protocol profiles. Because each protocol classifier probability result is independent, the sum of all protocol probabilities might exceed 100% as here. Autodesk have a probability of 9% because it only has the port heuristic positive result. Autodesk classifier result report report n/a (not available) because the analyzer does not have profile for this protocol.

3.2 Advanced Data

Advanced data are displayed from line 3 to 10. These data are gathered when signatures are successfully matched. The signature language allows to extract data from the payload such as the filename requested in the HTTP request (line 4 and 6). This is done as in Perl regular expression by adding capture parenthesis to the signature and adding the corresponding capture variable to one of the signature template. In the case of the HTTP filename, the variable was added to the filename template.

To provide a set of suitable templates for each type of information gathered *NetAnalyzer* uses four types of signatures: protocol, file, software, and user. For example the software version data is irrelevant when dealing with file analysis.

4 Probabilistic Identification Analysis

As exemplified in the above section, the analysis combines the result of multiple signatures matches along with identification heuristics. The protocol identification is said continuous because for each new packet, the session protocol is re-evaluated. As demonstrated in [17], protocol identification based on signature matching is more accurate than the identification based on classifier. Both of them are of course more reliable than the identification based on port heuristic. Thus the analysis uses the weighted arithmetic mean to reflect these different levels of accuracy. Accordingly the probability of a protocol $\mathbb{P}_x$ is computed as follows:

$$\mathbb{P}_x = \frac{\alpha \mathbb{H}_x + \beta \mathbb{C}_x + \gamma \mathbb{S}_x}{\alpha + \beta + \gamma}$$

where $\mathbb{H}_x$ is the protocol probability according to the port heuristic and α is its confidence coefficient. $\mathbb{H}_x$ is either equal to 0 or 100. $\mathbb{C}_x$ is the protocol probability according to the classifier heuristic and β is its confidence coefficient. $\mathbb{C}_x$ range from 0 to 100. Finally $\mathbb{S}$ is the protocol probability according to the payload analysis and σ is its reliability coefficient. In *NetAnalyzer*, we use the following values $\alpha = 1, \beta = 5, \sigma = 10$. They are consistent with [17] and work well in practice. The HTTP protocol probability of the example 1 is therefore:

$$\mathbb{P}_{http} = \frac{1 \times 0 + 5 \times 100 + 10 \times 100}{16} = 93.75$$

4.1 Classifier Probability

The classifier probability for a given protocol is computed by comparing each packet with a set of profiles. In its current implementation, *NetAnalyzer* only reports TCP protocol solely identified by profile if and only if there is no protocol identified by the payload analysis. ICMP and UDP identified protocols are always reported. The classifier probability is the ratio between successful matched packets and the total of packets:

$$\mathbb{C} = \frac{success}{total} * 100$$

In the above example (Figure 1), the 23 packets were successfully matched against HTTP profiles.

As in [41], server and client stream profiles are separated to improve classifier accuracy. This makes sense because often one stream is used to request data that are sent back by the other e.g the HTTP client request a file that is sent by the server. The classifier uses four discriminators: the packet size, the time elapsed since the last packet from the same stream, the time elapsed since the last packet from the other stream, and finally the entropy of the packet. The entropy use the Paninsky estimator [31] which is known to be efficient even against a small amount of data. As noted in [5], the set of profiles matched by a packet is related to it position in the stream: the stream first packets often contains request and authentication data. For example a POP3 session starts with a hello exchange followed by an authentication. That is why we keep a separate set of profiles for the first 10 packets of each stream and an aggregate one for the remaining packets. Thus each protocol has 22 profiles sets.

```
Ping:ICMP:2:2:64:64:995229:1004962:::7.54564:7.65728:
```

Fig. 2. ICMP ping packet 1 profile

For each discriminator, an upper bound and a lower bound are determined from the average mean and the standard deviation. If the packet value is between these bounds then it matches the discriminator. A packet matches a profile if it matches the four discriminators. We use a set of profiles rather than a single profile with very large bound to improve detection accuracy. We generate automatically profiles with smaller bound by using the standard k-means clustering [32] algorithm. We prevent profile over-fitness by limiting the number of profiles to 5 by set. During the clustering process, meaningless discriminators are removed. A discriminator is meaningless, if its standard deviation is too wide despite the clustering process. For example the entropy discriminator is removed if it has a standard deviation above 3. Figure 2 is an example of profile used in *NetAnalyzer*. The first field is the protocol name: Ping. The second field is the layer 3 protocol: ICMP. The third is the targeted stream: the number 1 is used for the server stream and 2 for client stream. The fourth field is the stream packet number. The exemple is a profile for the stream second packet. The rest of the profile are the four discriminators lower and upper bound: First the packet size (64 bytes), Secondly the time elapsed since the last packet from the same stream in microsecond (around 1 second here). Thirdly the time elapsed since the last packet from the opposite stream. Here it has been removed because it was meaningless.

Finally the packet entropy, which has to be between 7,54 and 7,65. Profiles are very efficient for covert channel detection as we will shown in Sect. 5

4.2   Payload Probability

Intuitively the payload analysis can be misled by tunneled session because the first signature matches belong to the tunnel protocol not the tunneled one. Hence the first thing to detect tunneled session is to perform continuous identification to identify the tunneled protocol. This is however not sufficient because the analyzer has also to figure out which one is the real protocol. This is done by taking into account the matches time line . Relying on the last match is not an option because it opens the door to injection attack that add an irrelevant payload at the end of the session. That is why the payload analysis gives to the latest matched signature a more significant weight than any previous match. To do so the technique uses a *weighted moving average* to compute each protocol probability:

$$\mathbb{P} = \frac{D_{x_i} \times n + D_{x-1} \times (n-1) + ... + D_{x_1} \times (1)}{n + (n-1) + ... + 1}$$

Where D_{x_i} is the confidence value associated to the signature matched in position i and n the global number of matches. The confidence value associated to a signature is by default 100. It can be however tweaked by the signature language option *confidence*. This is useful to decrease the confidence of a signature that is known to produce false positive. *NetAnalyzer* uses a specialized signature language to perform the payload analysis that allows to specify the signature confidence. For instance some of the signatures used to detect edonkey p2p traffic are known to produce from time to time false positive because they are quite short. However due to the edonkey protocol format they cant be improved, hence the only way to mitigate false positive is to reduce their confidence value.

5   Tunneled Session Detection

Tunneling a protocol into an another is a popular technique to bypass firewall restriction, hence a tunneled session is often a violation of the security policy.

As explained above (Sect. 4.2), the WMA *Weighted Moving Average* used to compute the payload analysis probability assigns a heavier weight to the most recent match. This is consistent with the fact that matches from the tunneled protocol will be reported after tunnel protocol ones.

```
(1) CONNECT irc.********.org:6667 HTTP/1.0
(2) HTTP/1.0 200 Connection established
(3) USER 0 0 0 ::
(4) NICK ****
(5) :irc.********.org 001 **** :Welcome to the **  **!0@xxx
```

Fig. 3. An exemple of a tunneled IRC session in a HTTP one

Figure 3 example is an IRC session tunneled into an HTTP one. The first two matches will identify the sessions as a HTTP one (line 1 and 2). If the identification process stops after the first match then the session is incorrectly identified as HTTP. With the continuous inspection, two matches for the IRC protocol will be reported (line 3 and 5). Hence the analysis ends up with four matches: two for HTTP and two for IRC. Because the analysis use a WMA, the two IRC matches will have an heavier weight, and therefore the session will be correctly identified as an IRC one. Payload probability scores are:

$$\mathbb{S}_{http} = \frac{100 + 200}{100 + 200 + 300 + 400} = 30\% \qquad \mathbb{S}_{irc} = \frac{300 + 400}{1000} = 70\%$$

Thereafter identification scores are:

$$\mathbb{HTTP} = \frac{100 \times 1 + 30 \times 10}{11} = 36,3\% \qquad \mathbb{IRC} = \frac{0 \times 1 + 70 \times 10}{11} = 66,6\%$$

Let's take a step further and imagine that the session is not a tunneled session but instead the download of the IRC RFC. In this context, the protocol identification has been misled. However, the previous computation does not take into account the probability given by the classifier. HTTP session and an IRC session exhibit very different profiles.

The figure 4 presents the packet size evolution for an IRC server stream whereas the figure 5 shows the packet size evolution for a HTTP server stream. In the HTTP stream, the packet size is constant after the first packet because when a file is sent to a client, the standard behavior is to maximize the throughput by sending the largest packet possible. This behavior is often referred as a TCP bulk transfers. Conversely because IRC is an interactive protocol the message size transmitted by the server to the client greatly fluctuate and never reach the maximum segment size. Hence every packet will be recognized by the classifier as HTTP one. Thereafter the complete protocol identification probability is:

$$\mathbb{HTTP} = \frac{100 \times 1 + 100 \times 5 + 30 \times 10}{16} = 56,25\% \qquad \mathbb{IRC} = \frac{0 \times 1 + 0 \times 5 + 70 \times 10}{11} = 43,75\%$$

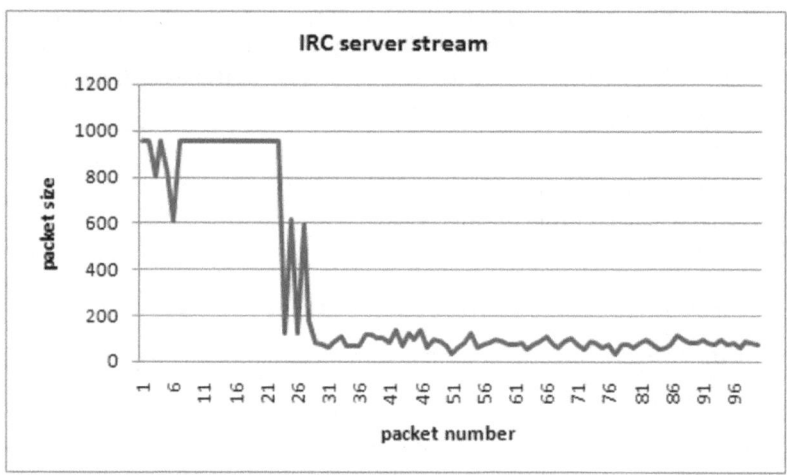

Fig. 4. An IRC server stream packet size evolution

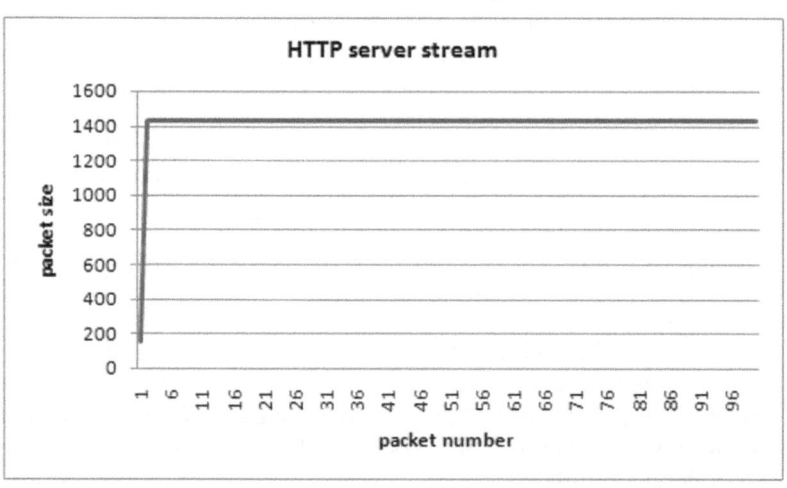

Fig. 5. An HTTP server stream packet size evolution

6   Covert Channel Detection

An other important problem is the detection of covert channel. Botnets and malware use them to hide there presence and bypass firewall restriction. The DDOS tool Stacheldraht [7] use an ICMP covert channel, and the backdoor Spotcom [40] use, to some extent a DNS one. ICMP covert channel are hard to detect because the RFC [33] state that the packet payload can be anything. DNS channel are also difficult because tunneled traffic hide in TXT record or other arbitrary length field. That is why covert channel are mainly uncovered by the traffic classifier. We present here how a ICMP tunnel is detected by the traffic classifier. For testing purpose, we have used the popular software PTunnel [38] to tunnel a SSH session into an ICMP covert channel. *NetAnalyzer* was able to detect it because the packets generated by the covert channel does not match ICMP ping profiles. Two discriminators exhibit very different behavior when it is a legitimate ping and when it is a covert channel. The first discriminator is the packet size: for a legitimate ping, the packet size is constant (diagram 6) whereas it fluctuates greatly for the covert channel (diagram 7).

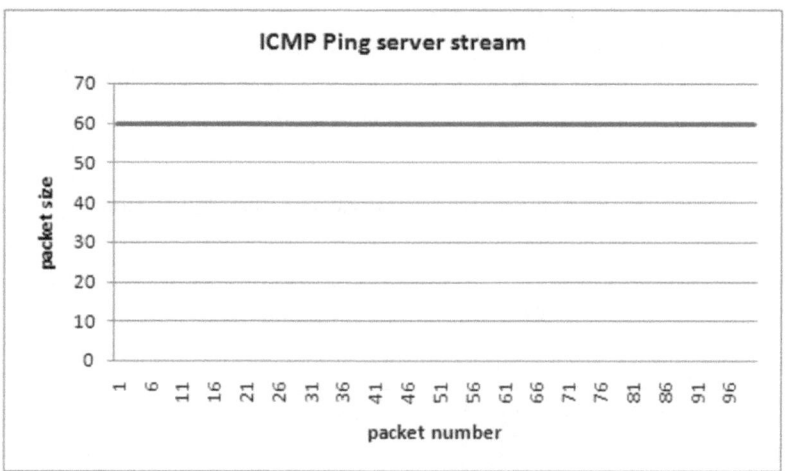

Fig. 6. An Icmp echo reply stream packet size evolution

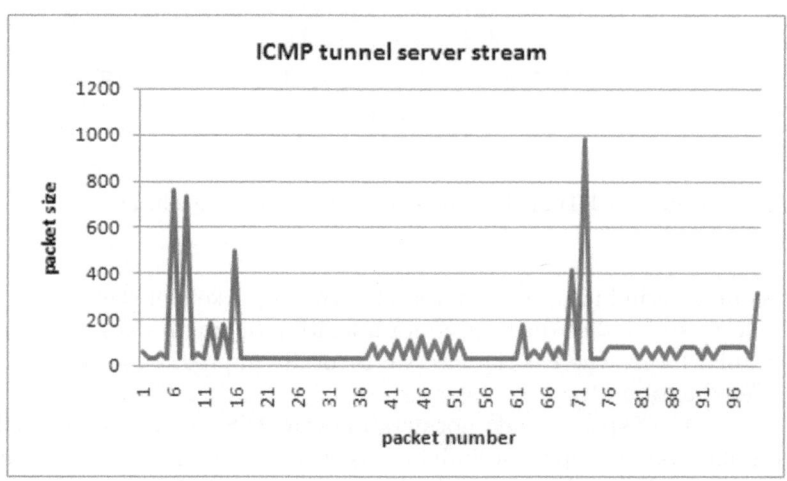

Fig. 7. A Ptunnel server stream packet size evolution

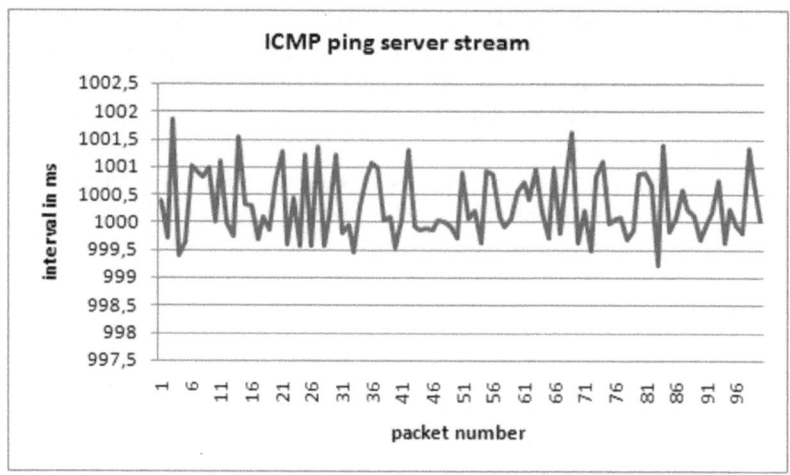

Fig. 8. An Echo reply stream packet interval evolution

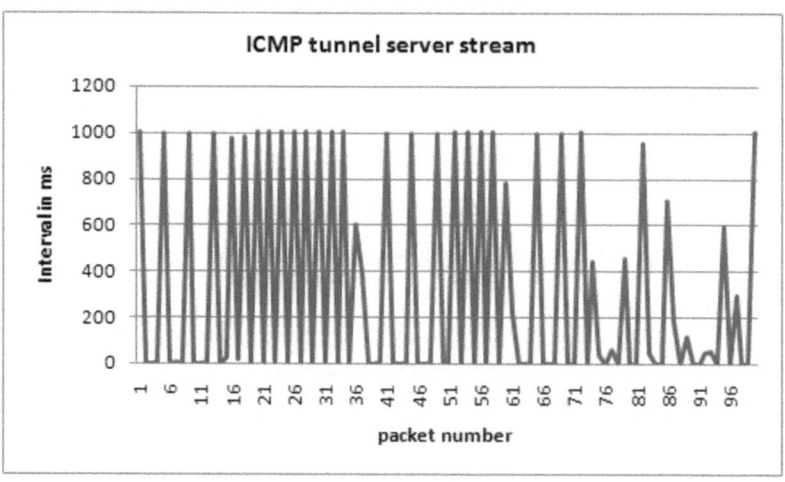

Fig. 9. A PTunnel server stream interval size evolution

The second discriminator is the interval between packets of the same stream. Values are very stable (See diagram 8) for a legitimate ping because requests are send on regular basis, whereas the interval is totally unpredictable for the covert channel (See diagram 9).

Due to the lack of space, we do not detail how a DNS channel is detected, but as noted in [20], the entropy discriminator is effective against it. Intuitively this is because the legitimate data uses a limited character set [26], that induce a low entropy whereas tunneled session have a high entropy. Traffic classifier is effective against covert channel, because it is very difficult to impersonate successfully every aspect of a given protocol. However some obfuscation techniques such as packet padding [13] or a combination of them [2] can be use to deceive it. In this case the only option is to find an other discriminator that will be not confused until a new obfuscation method is introduced.

7 File Masquerading Detection

The payload analysis allows *NetAnalyzer* to use content information to detect *File masquerading*. A file masquerading occurs when the extension of file does not reflect the content of the file *e.g* an *avi* file with a *.html* extension. This masquerading can mainly occurs for two reasons that deserve attention. The most common one, is when the masquerading is used to hide litigious or illegal file.

This a common practice used by many pornographic websites to deceive free hosting policy: they rename their photos or videos with textual extension such as .txt, to evade server log analysis that only rely on file extension. This technique is also commonly used by hackers to hide their illegal files on a compromised server. The second reason that leads to file masquerading is when a legitimate user makes a mistake. In this case detecting file masquerading is also important because it can prevents legitimate application to open properly the masqueraded file. A real world example of such masquerading is visible in figure 1: The first requested file is a gif masquerades as a jpeg file. This is not a serious issue as the browser will handle it, but nevertheless gif format is still proprietary.

8 Evaluation

To evaluate the effectiveness of our identification analysis presented in Sect 4, we run *NetAnalyzer* against a 8Gb trace of a residential network traffic. The goal was to evaluate the analysis ability to identify P2P sessions. We run two analysis : the first with only the port heuristic activated and the second with advanced analysis.

Analysis results are summarized in the following table:

Protocol	Port heuristic	Advanced inspection	Difference
HTTP	8589	9512	10,7%
BitTorrent	0	1504	n/a
Edonkey	1694	11249	564%
Unknown	16604	7564	-54.4 %
Others	3748	806	- 78,5%

When the identification use only the port heuristic, 54% of the traffic is reported as unknown which is coherent with the study [25]. Only 6% of the traffic is detected as Edonkey and none as BitTorrent (See figure 10). The traffic reported as "Other" include email and instant messaging traffic but also improbable protocols. This not surprising that port heuristic performs so poorly against Edonkey and BitTorrent traffic. Both of them use many obfuscation techniques.

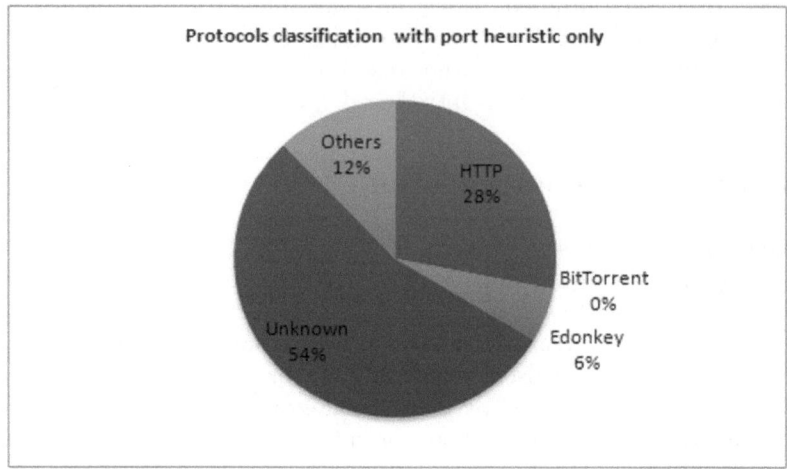

Fig. 10. Protocols classification based only on the port heuristic

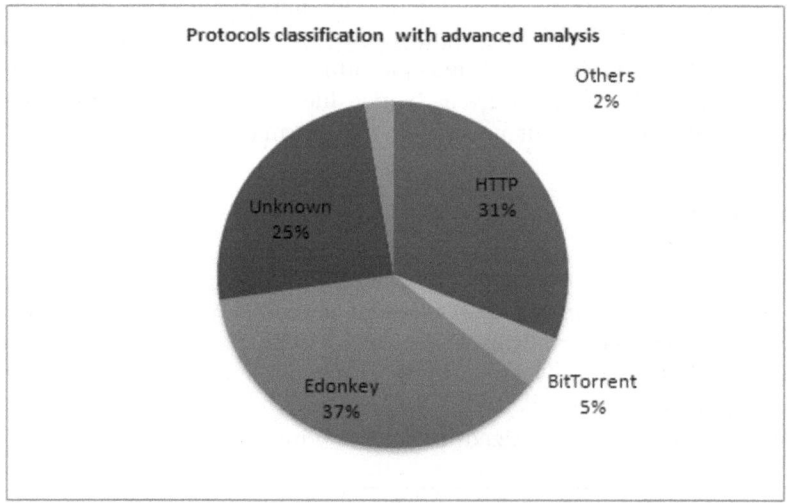

Fig. 11. Protocols classification that uses the advanced analysis

The popular client for the Edonkey network Emule [34] enable by default obfuscation techniques, such as padding, since version 0.47b . Popular BitTorrent client such Azureus [1] recommends to not use standard port and can use a very effective obfuscation scheme called "Message Encryption Stream" [2] based en public key cryptography and payload randomization. This scheme, also used by μTorrent, is designed "*to provide a completely random-looking header and (optionally) payload to avoid passive protocol identification and traffic shaping.*" In particular it can use RC4 and DH to encrypt the packet payload.

When the advanced analysis is enabled, the protocol classification accuracy improves drastically (See figure 11). This time the traffic classification shows that more than 60% of the traffic is in fact P2P traffic. It also reduce the number of unknown traffic by 54.4%, and to increase the number of Edonkey session identified by 564%. The advanced analysis is able to uncover BitTorrent traffic, something that the heuristic based analysis was unable to achieve. The number of others protocol also shrinks down by 78,5%: the advanced analysis was able to reduce the number of sessions incorrectly identified. Even if the protocol classification is still not perfect, as 25% of the traffic remains unclassified, the advanced analysis as been proved effective to improve significantly the protocol classification when hard to classify protocols are used.

9   Conclusion

The main contribution of this paper was a probabilistic identification analysis and its implementation in a passive network monitor called *NetAnalyzer*. We have shown that it overcomes the limitations of traditional port-based protocol identification when dealing with hard to classify protocol. We also have shown that the analysis is able to deal with tunneled and covert channel. A Future work is to introduce specific discriminator for P2P to improve further more the identification. The UDP and TCP connection pairing discriminator proposed in [21] seems promising.

References

1. Azureus. Increase download speed recommandation,
   http://www.azureuswiki.com/index.php/increase_download_speed
2. Azureus. Message stream encryption,
   http://www.azureuswiki.com/index.php/message_stream_encryption
3. Babachiz. Managing peer-to-peer traffic with cisco service control technology. Technical report, CISCO (Febuary 2005)
4. Baset, S., Schulzrinne, H.: An analysis of the skype peer-to-peer internet telephony protocol. In: IEEE infocom 2006 (2006)
5. Bernaille, L., Teixeira, R., Akodkenou, I., Soule, A., Salamatian, K.: Traffic classification on the fly. SIGCOMM Comput. Commun. Rev. 36(2), 23–26 (2006)
6. Bursztein, E.: Netanalyzer homepage (2006), http://www.netqi.org
7. Cheng, G.: Analysis on ddos tool stacheldraht v1.666,
   http://www.sans.org/resources/malwarefaq/stacheldraht.php
8. Chung, S.P., Mok, A.K.: Allergy attack against automatic signature generation. In: RAID Recent Advances in Intrusion Detection (2006)
9. Constantinou, F., Mavrommatis, P.: Identifying known and unknown peer-to-peer traffic. In: NCA 2006: Proceedings of the Fifth IEEE International Symposium on Network Computing and Applications, Washington, DC, USA, pp. 93–102. IEEE Computer Society, Los Alamitos (2006)
10. Deri, L., Suin, S.: Ntop: Beyond ping and traceroute. In: Stadler, R., Stiller, B. (eds.) DSOM 1999. LNCS, vol. 1700, pp. 271–283. Springer, Heidelberg (1999)

11. Dreger, H., Feldmann, A., Mai, M., Paxson, V., Sommer, R.: Dynamic application-layer protocol analysis for network intrusion detection. In: Usenix (2006)
12. Dvoinikov, D.: Htthost (http proxy), http://www.htthost.com/htthost.boa
13. Fu, X., Graham, B., Bettati, R., Zhao, W.: On effectiveness of link padding for statistical traffic analysis attacks. In: ICDCS 2003: Proceedings of the 23rd International Conference on Distributed Computing Systems, Washington, DC, USA, p. 340. IEEE Computer Society, Los Alamitos (2003)
14. Fyodor. Nmap: free open source utility for network exploration or security auditing, http://www.insecure.org/nmap/
15. Del Corso, F.R.G.M.: Ranking a stream of news. In: 14th international conference on World Wide Web, p. 97 (2005)
16. Gummadi, P.K., Saroiu, S., Gribble, S.D.: A measurement study of napster and gnutella as examples of peer-to-peer file sharing systems. SIGCOMM Comput. Commun. Rev. 32(1), 82–82 (2002)
17. Haffner, P., Sen, S., Spatscheck, O., Wang, D.: Acas: automated construction of application signatures. In: MineNet 2005: Proceeding of the 2005 ACM SIGCOMM workshop on Mining network data, pp. 197–202. ACM Press, New York (2005)
18. IANA. Matrix for protocol parameter assignment/registration procedures, http://www.iana.org/numbers.html
19. Java, J.: Iptraf (ip network monitoring software), http://iptraf.seul.org/
20. Kaminsky, D.: Attacking distributed systems the dns case study. In: Black Hat Europe (2005)
21. Karagiannis, T., Broido, A., Faloutsos, M., claffy, k.: Transport layer identification of p2p traffic. In: IMC 2004: Proceedings of the 4th ACM SIGCOMM conference on Internet measurement, pp. 121–134. ACM Press, New York (2004)
22. Kreibich, C., Crowncroft, J.: Creating intrusion detection signatures using honeypot. In: Second Workshop on Hot Topics in Networks (November 2003)
23. Kruegel, C., Vigna, G.: Anomaly Detection of Web-based Attacks. In: Proceedings of the 10th ACM Conference on Computer and Communication Security (CCS 2003), Washington, DC, October 2003, pp. 251–261. ACM Press, New York (2003)
24. Li, Z., Sanghi, M., Chavez, B., Chen, Y., Kao, M.-Y.: Hamsa: Fast signature generation for zero-day polymorphic worms with provable attack resilience. In: IEEE Symposium on Security and Privacy (May 2006)
25. Madhukar, A., Williamson, C.: A longitudinal study of p2p traffic classification. In: MASCOTS 2006: Proceedings of the 14th IEEE International Symposium on Modeling, Analysis, and Simulation, Washington, DC, USA, pp. 179–188. IEEE Computer Society, Los Alamitos (2006)
26. Mockapetris, P.: Rfc1035: Domain names - implementation and specification. Technical report, Network Working Group (1987)
27. Moore, A.W., Zuev, D.: Internet traffic classification using bayesian analysis techniques. In: SIGMETRICS 2005: Proceedings of the 2005 ACM SIGMETRICS international conference on Measurement and modeling of computer systems, pp. 50–60. ACM Press, New York (2005)
28. Newsome, J., Karp, B., Song, D.: Polygraph: Automatically generating signatures for polymorphic worms. In: IEEE Symposium on Security and Privacy (May 2005)
29. Ohzahata, S., Hagiwara, Y., Terada, M., Kawashima, K.: A traffic identification method and evaluations for a pure p2p application. In: Dovrolis, C. (ed.) PAM 2005. LNCS, vol. 3431, pp. 55–68. Springer, Heidelberg (2005)
30. Page, L., Brin, S., Motwani, R., Winograd, T.: The pagerank citation ranking: Bringing order to the web. Technical report, Stanford Digital Library Technologies Project (1998)

31. Paninski, L.: Estimation of entropy and mutual information. Neural Comput 15(6), 1191–1253 (2003)
32. Pelleg, D., Moore, A.W.: X-means: Extending k-means with efficient estimation of the number of clusters. In: ICML, pp. 727–734 (2000)
33. Postel, J.: Rfc792: Internet control message protocol. Technical report, Network Working Group (1981)
34. Project, E.: http://www.emule-project.net/
35. Quadong. Linux layer 7 packet classifier, http://sourceforge.net/projects/l7-filter/
36. Shannon, C.E.: Prediction and entropy of printed english. Bell System Technical Journal (1951)
37. Singh, S., Estan, C., Varghese, G., Savage, S.: Automated worm fingerprinting. In: Operating Systems Design and Implementation (December 2004)
38. Stødle, D.: Ping tunnel
39. Tsaparas, P.: Link Analysis Ranking. PhD thesis, University of Toronto (2004)
40. Turkulainen, J.: Backdoor spotcom analysis, http://www.securiteam.com/securityreviews/6z00n208uc.html
41. Wang, K., Cretu, G., Stolfo, S.J.: Anomalous payload-based worm detection and signature generation. In: Recent Advance in Intrusion Detection (2005)
42. Wright, C., Monrose, F., Masson, G.M.: Hmm profiles for network traffic classification. In: VizSEC/DMSEC 2004: Proceedings of the 2004 ACM workshop on Visualization and data mining for computer security, pp. 9–15. ACM Press, New York (2004)

A Self-certified and Sybil-Free Framework for Secure Digital Identity Domain Buildup

Christer Andersson[1], Markulf Kohlweiss[2], Leonardo A. Martucci[1], and Andriy Panchenko[3]

[1] Karlstads Universitet, Department of Computer Science
Universitetsgatan 2, 651-88 Karlstad, Sweden
[2] Katholieke Universiteit Leuven, ESAT/COSIC
Kasteelpark Arenberg, 10 B-3001 Leuven-Heverlee, Belgium
[3] RWTH Aachen University, Department of Computer Science
Informatik IV, Ahornstr. 55, D-52074 Aachen, Germany
{christer.andersson, leonardo.martucci}@kau.se,
markulf.kohlweiss@esat.kuleuven.be, panchenko@cs.rwth-aachen.de

Abstract. An attacker who can control arbitrarily many user identities can break the security properties of most conceivable systems. This is called a "Sybil attack". We present a solution to this problem that does not require online communication with a trusted third party and that in addition preserves the privacy of honest users. Given an initial so-called Sybil-free identity domain, our proposal can be used for deriving Sybil-free unlinkable pseudonyms associated with other identity domains. The pseudonyms are self-certified and computed by the users themselves from their cryptographic long-term identities.

1 Introduction

Today, users often need to communicate and cooperate in networked environments. Virtual / Online communities, peer-to-peer systems, and applications for anonymous communication are only some prominent examples. Often, these systems depend on a majority of users being honest for tasks like voting in virtual community, reputation computation, Byzantine fault tolerance, or traffic mixing. Unless such systems implement expensive countermeasures they fall prey to the Sybil Attack [17], which entails a single attacker controlling arbitrarily many user accounts (called Sybil identities). Moreover, both identity certificates and the most advanced non-centralized Sybil defence mechanisms [23, 29] are privacy-invasive.

This paper defines *identity domains* as domains that uniquely specifies the context in which a set of identifiers is used. The purpose of a domain is to build an anonymity set i.e. a set of identifiers within an user is not identifiable. The context may include validity time, location, application, or other parameters. A secure identity domain should provide a Sybil-free environments (i.e., absent of Sybil identities) in which applications can be deployed. This paper shows

J.A. Onieva et al. (Eds.): WISTP 2008, LNCS 5019, pp. 64–77, 2008.

how, given one initial Sybil-free identity domain, we can propagate the Sybil-freeness to arbitrary many identity domains. In every identity domain each user is known under a different and unique pseudonym, and further there is no need of the continuous involvement of a Trusted Third Party (TTP). Access to a Certificate Authority (CA) is required only for the bootstrapping of a Sybil-free domain[1].

We call our solution *self-certified Sybil-free pseudonyms*[2]. These pseudonyms do not depend on the continuous availability of a TTP and, they are fully unlinkable. This is achieved using a self-certification mechanism: self-certified Sybil-free pseudonyms use concepts such as anonymous credentials and group signatures to enable the generation of an arbitrary number of anonymous certificates – however, only one certificate per identity domain and user identity. Access to the certificate authority (CA) is required only for acquiring the membership certificated from which the self-certified pseudonyms are derived from. Our solution can be seen as a framework that enables privacy-enhanced and Sybil-resistant buildup of user groups. We use periodic n-times spendable e-tokens [9] as a base for the instantiation, although there are also other cryptographic primitives that can be used to create such pseudonyms.

A user that wants to participate in the system first enrolls with the CA to acquire exactly one *membership certificate*. Thereby, we establish the initial Sybil-free identity domain. Using the certificate, the user can create one *self-certified pseudonym* per newly created identity domain. Membership certificates can be used for issuing pseudonyms for arbitrarily many identity domains, but the pseudonyms are only valid within the domain they were issued for. Further, pseudonyms issued for different identity domains are mutually unlinkable. Specifically, they cannot be linked to the underlying membership certificate even by the CA itself. The self-generated pseudonym certificates that come with self-certified pseudonyms provide three main functions: (*i*) binding of a freshly generated public key to the pseudonym (as with identity certificates); (*ii*) verification of the pseudonym and the binding; and (*iii*) disclosure of the user identity and revocation of her certificates, if the same membership certificate is used to create two different pseudonym certificates for the same domain.

This paper is organized as follows. The assumptions are presented in Section 2. Section 3 discusses applications that can benefit from our solution. The underlying cryptographic mechanism for our solution – n-spendable e-tokens – are introduced in Section 4. Then, the instantiation of our Sybil-free self-certified based on the e-tokens is described in Section 5. Finally, Section 6 concludes the paper.

---

[1] Identity domains can be constructed assuming continuous availability of a TTP. The problems with this approach are that the TTP can link all pseudonyms to the issuing user and the availability requirements on the TTP.

[2] Self-certified pseudonyms are also discussed in [2]. Whereas this paper presents a detailed description on their construction, the main focus of [2] is on their application in mobile ad hoc networks.

2   Assumptions

We assume that: (i) the CA is capable of establishing the initial Sybil-free domain; (ii) identity domain identifiers ctx are unique; and, (iii) devices are capable of performing the necessary cryptographic functions. Regarding the attacker model, the attacker has two main goals: (i) deploy a Syibil attack in a given identity domain and (ii) identify a relationship between two pseudonyms generated for different identity domains to find out if those pseudonyms belong to a same user. We assume that attackers are able to eavesdrop all network communication, but each attacker has at most one membership certificate $cert_{u}$.

A serious challenge regarding an eventual real life implementation of our system is the realization of the initial Sybil-free domains. Yet, this assumption is not exclusive for our scheme. In fact, for any scheme based on certificates, regardless of whether a fully distributed or a completely distributed security model is used (or something in between, such as a threshold scheme), some entity (or a cluster of several entities) *must* be trusted not to hand out more that one credential per identity (at least, it must be made very costly to obtain several credentials).

3   The Need for Sybil-Free Applications

The number of applications where a group of users interact electronically is endless: numerous instant messaging applications, chat rooms, forums and e-commerce platforms are only a few examples of widely used applications. Often, such applications allow users to slip into different roles, and behave accordingly. However, with growing size and sophistication of such communities and applications, the amount of required administration tasks grows: misbehaving users need to be excluded, user contributions need to be evaluated based on user reputation, and work tasks need to be distributed. In short, such applications and communities develop their own social dynamics, and there is a need to make decision processes work in a more automated way. For instance, such decisions could be based on majority voting, seniority, or reputation.

Truly anonymous or pseudonymous applications are currently debated, partly because they can enable misbehaving users to create social problems within their communities. Although these users can be banned from such applications, it is often easy for the wrongdoers to simply re-register using a new name. To change IP addresses using proxies or similar techniques is enough to thwart most existing countermeasures. Reputation systems also break under such an attack as users can register multiple times to collaboratively increase the reputation of all of their pseudonyms. Further, they can manipulate the allocation of resources and the distribution of work. Evil users can also choose names similar to other users to abuse their reputation. Finally, users that control multiple identities can more easily spread rumors and influence voting results to their own advantage.

Nonetheless, this separation between real world identities and different virtual worlds that allows the support of pseudonymous and anonymous users is a valued feature. As networks are "unforgetful" and may log and remember a close to infinite number of network interactions, this separation decreases the privacy risks associated with interacting in a computer network. Many scientific papers have been dedicated to various types of pseudonymity [5, 19, 4] or the graceful degradation of anonymity towards full identification [1] using existing approaches.

Numerous applications would benefit from having such a basic building block in place (even if the Sybil protection of the initial domain would only be approximate):

- *Resilient systems* require a majority of users to be honest in order to achieve Byzantine fault tolerance;
- *Peer-to-peer systems* need to manage reputation, some of which may rely on dummy e-currencies and distributed double-spending detection;
- *Online communities.* On platforms like eBay, a protection against self-ranking can be provided. Furthermore, if a user deletes his account and joins again (to get rid of "bad" reputation), both actions can be linked. Dating communities can be protected in the way that only one profile can be posted per physical user. Some online forums provide automatic banning of users if some fraction of the users vote for it. By using surveys, it is possible to make sure that a disjoint set of people was questioned;
- *Online multi-user games* need to be protected against cheaters. Privacy-friendly subscriptions with protection against sharing can be provided. Further, exclusion of bots from the game can be achieved;
- *Anonymous communication systems* require some portion of the users to be honest. They assume that nodes on the path between the sender and the receiver belong to different entities and do not cooperate. Otherwise, anonymity can be easily compromised. We investigate an example of such a system in [2].

We expect Sybil-free self-certified pseudonyms to be used in admission control schemes [21, 25, 26] to aid applications, such as those discussed in this section to manage anonymous or pseudonymous users in a secure and privacy-respecting manner. Privacy-friendly admission control allows to create and manage identity domains comprised of several parallel and unlinkable identity domains. Thus, a user can be part of multiple identity domains simultaneously (e. g. different online communities) and keep the identities used in different domains unlinkable.

4   Preliminaries: k-Spendable E-Tokens

We use a special signature scheme for creating pseudonym certificates: Camenisch *et al.* have proposed a protocol for periodically spendable e-tokens [9]. In their scenario, sensors spend an e-token whenever they report some data. Yet, it is only possible to compute k different e-tokens per time period. Consequently, sensors can file at most k reports per time period anonymously. Otherwise the sensors have to spend some e-token twice, which allows everyone to compute the sensor's identity from these two e-token show transcripts.

The e-token based signature scheme consists of the algorithms IKg, UKg, $Obtain$, $Issue$, $Sign$, $Verify$, $Identify$, and $Revoke$. These algorithms are executed by the issuer $\mathcal{I}$ of e-token dispensers, the user $\mathcal{U}$, and the signature verifiers:

- $IKg(1^k)$ and $UKg(1^k, pk_{\mathcal{I}})$ – creates the issuers key pair $(pk_{\mathcal{I}}, sk_{\mathcal{I}})$ and the user's key pair $(pk_{\mathcal{U}}, sk_{\mathcal{U}})$, respectively. The value k is the security parameter;
- $Obtain(pk_{\mathcal{I}}, sk_{\mathcal{U}}) \leftrightarrow Issue(pk_{\mathcal{U}}, sk_{\mathcal{I}})$ – at the end of this protocol between a user and the e-token issuer, the user obtains an e-token dispenser D that can be used to create one e-token based signature per ctx. I stores $pk_{\mathcal{U}}$ and revocation information r_D under the user's identity;
- $Sign(m, D, pk_{\mathcal{I}}, ctx)$ – shows an e-token from dispenser D in context ctx to sign a message m. The outputs are a token serial number (TSN) S, a transcript τ, and an updated e-token dispenser D';
- $Verify(m, S, \tau, pk_{\mathcal{I}}, ctx)$ – checks that S and τ were created by a valid dispenser D to sign a message m in context ctx;
- $Identify(pk_{\mathcal{I}}, S, \tau, \tau', m, m')$ – given two records (S, τ) and (S, τ') created by a dispenser D when signing m and m', $m \neq m'$, respectively, $Identify$ computes the public key $pk_{\mathcal{U}}$ of the owner of D;
- $Revoke(sk_{\mathcal{I}}, pk_{\mathcal{I}}, r_D)$ – takes as input $sk_{\mathcal{I}}$ and $pk_{\mathcal{I}}$ and the revocation information r_D that corresponds to a particular user (see $Obtain$). It outputs an updated issuer public key $pk_{\mathcal{I}}'$. In the rest of the paper, we assume that all parties use the most up-to-date issuer key for signing and verification.

4.1   Cryptographic Related Work

Different cryptographic systems can be used to create unlinkable and unique pseudonyms. As long as the identification of "double-spent" pseudonyms is not an issue, such pseudonyms can be realized based on the so-called epoch number of direct anonymous attestation [8]. Schemes that support identification were presented in [9] and [15]. By binding a different tag to every identity domain, k-times anonymous authentication [28] can be used to create unique pseudonyms. Our scheme uses the cryptographic techniques of Camenisch et al. [9] (i.e., e-tokens), but can be seen as a more general systems framework that could also be instantiated using other cryptographic techniques.

4.2   Realization of the Cryptographic Algorithms

Briefly, the above functionality can be realized as follows. The issuer and the user both generate key pairs. Let the user's key pair be $(pk_{\mathcal{U}}, sk_{\mathcal{U}})$, where $pk_{\mathcal{U}} = g^{sk_{\mathcal{U}}}$ and g generates a group $\mathbb{G}$ of known order. The issuer's key pair is used for creating and verifying CL signatures. We use a PRF f_s whose range is the group $\mathbb{G}$. Using $Obtain$, the user interacts with the issuer running the $Issue$ algorithm and obtains an e-token dispenser D that allows her to show one e-tokens per context. The dispenser D is comprised of seed s for the PRF f_s, the user's secret key sk_U, and the issuer's CL signature on $(s, sk_{\mathcal{U}})$. CL signatures are used to prevent the issuer from learning anything about s or $sk_{\mathcal{U}}$. Moreover, the dispenser D is revoked by revoking the corresponding CL signature [11]. In the *Sign* algorithm, the user shows her token for the context ctx: she releases a serial number $S = f_s(0\|ctx)$, a double-show tag $E = pk_{\mathcal{U}} \cdot f_s(1\|ctx)^{h(m)}$, and using the Fiat-Shamir heuristic [18] creates a non-interactive ZK proof σ that (S, E) correspond to a valid dispenser for context ctx (i.e., the user proves in zero-knowledge that S and E were properly formed from values $(s, sk_{\mathcal{U}})$ signed by the issuer). To sign message m, m is hashed into the challenge together with the first message and the public parameters of the proof. The transcript τ contains both E and σ. An e-token is verified by checking the non-interactive proof.

Unlinkability and Identification. As f_s is a pseudo-random function, and all proof protocols are zero-knowledge, it is computationally infeasible to link the resulting e-token to the user, the dispenser D, or any other e-tokens corresponding to D. If a user shows two e-tokens in the same context to authenticate two messages m and m', then both e-tokens *must* use the same serial number. The issuer can easily detect the violation and compute $pk_{\mathcal{U}}$ from the two double-show tags, $E = pk_{\mathcal{U}} \cdot f_s(1\|ctx)^{h(m)}$ and $E' = pk_{\mathcal{U}} \cdot f_s(1\|ctx)^{h(m')}$. From the equations above, $f_s(1\|ctx) = (E/E')^{(h(m)-h(m'))^{-1}}$ and $pk_{\mathcal{U}} = E/f_s(1\|ctx)^{h(m)}$. For a more detailed security analysis, we refer the reader to [9].

4.3 Cryptographic Details

This writeup is based on a similar writeup for compact e-cash [10]. See the Appendix for more details about the cryptographic primitives used in the writeup. To provide the full protocols for e-token signatures, we provide details about the CL signature scheme used by the issuer. Let QR_n denote the set of quadratic residues modulo n. Let Z, U, V be elements of QR_n that are part of the public key of the issuer. Let l_n denote the number of bits of the issuer's RSA modulus n and let l_o be a security parameter controlling the statistically zero knowledge property of the proof protocol as well as the statistically hiding property of the commitment schemes we use. A signature of the issuer on the seed (message) s consists of the values (Q, e, v), where $e \in \{2^{l_e} - 2^{l_{e'}}, 2^{l_e} + 2^{l_{e'}}\}$ is a random prime, $v \in \{0, 1\}^{l_n + l_o}$ is a random integer, and $Q \in \langle U \rangle \subset QR_n$, such that the following holds:

$$Z \equiv Q^e V^v U^s \ (mod \ n).$$

The issuer does not learn s when issuing this signature. Rather, the issuer and user run a two-party protocol where the output of the user will be (Q, e, v). If the issuer signs a block of messages at once, say (u, s), we replace V by (V_1, V_2) in the public key. The signature still consists of values (Q, e, v) such that

$$Z \equiv Q^e V_1^u V_2^s U^s \ (mod \ n).$$

We now describe how the user creates an e-token signature in more detail. Recall that the user has obtained from the issuer a signature (Q, e, v) on u and s.

1. The user computes the serial number $S = g^{1/(s+0\|ctx)}$ and a security tag $E = pk_{\mathcal{U}} g^{H(m)/(s+1\|ctx)}$

2. The user chooses a random r_B and computes the commitments $B = \mathbf{g}^u \mathbf{h}^{r_B}$.
3. The user chooses a random r and computes $Q' := QU^r$. Note that $(Q', e, v + r)$ is also a valid signature on the message u and s but that Q' and Q are statistically independent.
4. The user computes the following signature proof of knowledge:

$$
\begin{aligned}
\sigma = SPK\{(\alpha, \mu, \gamma, \zeta, \epsilon, \rho_1, \rho_2) : \\
Z = \pm Q'^{\epsilon} V_1^{\mu} V_2^{\gamma} U^{\zeta} \wedge \\
g = S^{\gamma} S^{0\|ctx} \wedge B = \mathbf{g}^{\mu} \mathbf{h}^{\rho_2} \wedge \\
1 = B^{\gamma} B^{1\|ctx} (1/\mathbf{g})^{\alpha} \mathbf{h}^{\rho_1} \wedge \\
g^{H(m)} = E^{\gamma} E^{1\|ctx} (1/g)^{\alpha} \wedge \\
\gamma \in \{0,1\}^{l_m + l_o + l_n + 2} \wedge \\
(\epsilon - 2^{l_e}) \in \{0,1\}^{l_{e'} + l_o + l_n + 1} \\
\}(Z, V_1, V_2, U, \mathbf{g}, \mathbf{h}, n, g, S, E, B, m)
\end{aligned}
\tag{1}
$$

The parts of the proof related to the CL signature are done over QR_n while the proofs for S and E are done in $\mathbb{G}$. Elements g and h are generators of $\mathbb{G}$; $\mathbf{g}$ and $\mathbf{h}$ are generators of QR_n.

5   Self-certified Sybil-Free Pseudonyms

In this section we describe how self-certified Sybil-free pseudonyms can be constructed and used.

5.1   Instantiation Based on E-Token Signatures

In this section, we describe how to implement self-certified Sybil-free pseudonyms by using e-token signatures as a base. The pseudonym certificates $cert_{(\mathcal{U}, ctx)}$ that come with the self-certified pseudonyms provide three main functions: (i) the binding of a freshly generated public key to the pseudonym (as with identity certificates); (ii) the verification of the pseudonym and the binding, and; (iii) the disclosure of the user identity and revocation of her certificates should the same membership certificate be used to create two different pseudonym certificates for the same identity domain.

While k-spendable e-tokens provide the necessary main functionality for fulfilling our requirements, we adapt their solution in several ways: (i) while their show protocol is interactive we require non-interactive publicly verifiable shows for signature verification; (ii) we bind a temporal public key to the e-token show – the public key is the message that is signed; (iii) instead of time periods we limit the number of generated e-tokens per signing context – while a context has a validity period, it may also have a name and other characteristics; and, (iv) we use a version optimized for $k = 1$. The first two properties are obtained by applying the Fiat-Shamir heuristic [18], a cryptographic trick that turns certain

Table 1. A summary of the notation used on the conceptual and the solution level

Conceptual Level	Solution Level
membership certificate $cert_\mathcal{U}$	dispenser D
pseudonym certificate $cert_{(\mathcal{U},ctx)}$	transcript τ
pseudo-random pseudonym $P_{(\mathcal{U},ctx)}$	serial number S
domain identifier, context descriptor ctx	

interactive identification protocols into signature schemes. Instead of a time period t, we use an arbitrary context identifier ctx. The value ctx can be seen as identifying the context in which a signer is allowed to sign only once.

The interaction model of our proposal consists of two "phases", one *enrollment* phase in which an initial Sybil-free identity domain is established, and one *identity domain buildup and use* phase where users create and maintain identity domains derived from the original identity domain. See Table 1 for a summary of our notation for the conceptual and the solution level, respectively, and the entities of our system and the roles they may assume are listed in Table 2.

Table 2. A summary of the system entities and their respective roles

Entities	Trusted	Possible Roles
Certificate Authority	Yes	Issuer
User	No	User, verifier, domain controller

Enrollment. This phase involves several users and one issuer – the certificate authority $\mathcal{I}$. Initially $\mathcal{I}$ generates an e-token issuing key pair $(pk_\mathcal{I}, sk_\mathcal{I})$ using IKg. To enroll, a user $\mathcal{U}$ creates a membership key pair $(pk_\mathcal{U}, sk_\mathcal{U})$ using UKg. She transfers $pk_\mathcal{U}$ to $\mathcal{I}$ and authenticates under her identity for the Sybil-free identity space. In turn, $\mathcal{U}$ and $\mathcal{I}$ interact using the $Obtain(pk_\mathcal{I}, sk_\mathcal{U}) \leftrightarrow Issue(pk_\mathcal{U}, sk_\mathcal{I})$ protocol. In this way $\mathcal{U}$ obtains an e-token dispenser D. It is used as her membership certificate $cert_\mathcal{U}$ (see Table 1).

Identity Domain Buildup and Use. In this phase, users collectively buildup and participate in identity domains. It consists of three subphases, during which a subset of the users may take the roles of domain controller and / or verifier.

- *Identity domain context creation.* To create a context for an identity domain, a domain controller publishes a domain identifier ctx. As a heuristic, a long-lived ctx should follow some kind of URI-like (Uniform Resource Identifier) scheme and a short-lived ctx should include the identity domain's validity time. The uniqueness of the domain identifiers used by a user $\mathcal{U}$ can be guaranteed under three conditions: (*i*) $\mathcal{U}$ never turns back her clock; (*ii*) $\mathcal{U}$ keeps a list of all the domain identifiers she has used, and removes records from the list only if the corresponding identity domains have expired; (*iii*) $\mathcal{U}$

only joins domains that have not yet expired and whose domain identifiers are not already on her list.

In addition, ctx may contain the name of the domain, the public key of the domain controller, or even a contract that all of the users who join the domain should agree on. From a practical point of view, there is no hard limit on the size of ctx. It can be hashed down to a constant size value before being used in the cryptographic algorithms. Appending the hash to the validity time makes the uniqueness of ctx independent from the collision resistance of the hash function.

As the identity domain controller does not need to be trusted, any user (or several users) could perform this role. Besides publishing ctx, the identity domain controller will often be responsible for distributing pseudonym certificates. The user that controls the identity domain controller can also participate in the domain, issuing its own pseudonym certificate.

- *Pseudonym certificate creation and verification.* Registration at an identity domain is done using the triplet $(pk_{(\mathcal{U},ctx)}, P_{(\mathcal{U},ctx)}, cert_{(\mathcal{U},ctx)})$, generated as follows: a user $\mathcal{U}$ with a membership certificate $cert_\mathcal{U}$ wants to certify a new application specific and hitherto uncertified public / private key pair, which we will from now on call $(pk_{(\mathcal{U},ctx)}, sk_{(\mathcal{U},ctx)})$. She creates a pseudo-random pseudonym $P_{(\mathcal{U},ctx)}$ for a given ctx using the e-token to sign $pk_{(\mathcal{U},ctx)}$. The $Sign(pk_{(\mathcal{U},ctx)}, cert_\mathcal{U}, pk_\mathcal{I}, ctx)$ algorithm outputs an e-token-based signature (S, τ). $\mathcal{U}$ uses the e-token's serial number S as her pseudo-random pseudonym $P_{(\mathcal{U},ctx)}$, and the transcript τ as her pseudonym certificate $cert_{(\mathcal{U},ctx)}$ (see Table 1). Hence, the domain controller cannot prevent a qualified user (i. e. an user $\mathcal{U}$ with a membership certificate $cert_\mathcal{U}$) that knows ctx to join the domain.

  Any user can now verify the correctness of $cert_{(\mathcal{U},ctx)}$ using $Verify(pk_{(\mathcal{U},ctx)}, P_{(\mathcal{U},ctx)}, cert_{(\mathcal{U},ctx)}, pk_\mathcal{I}, ctx)$. Afterwards, the uniqueness of the pseudonym can be checked by comparing $P_{(\mathcal{U},ctx)}$ with the pseudonyms of the other certificates for this domain by executing $Identify$. These verifications can be done by any node that is part of the domain, at any choosen time.

- *Misuser identification and revocation.* By executing $Identify$, it is possible to extract the membership public key $pk_\mathcal{U}$ of a user from two pseudonym registrations $(pk_{(\mathcal{U},ctx)}, P_{(\mathcal{U},ctx)}, cert_{(\mathcal{U},ctx)})$ and $(pk'_{(\mathcal{U},ctx)}, P'_{(\mathcal{U},ctx)}, cert'_{(\mathcal{U},ctx)})$ if $P_{(\mathcal{U},ctx)} = P'_{(\mathcal{U},ctx)}$ (assured by the system) and $pk_{(\mathcal{U},ctx)} \neq pk'_{(\mathcal{U},ctx)}$. $Identify(pk_T, P_{(\mathcal{U},ctx)}, cert_{(\mathcal{U},ctx)}, cert'_{(\mathcal{U},ctx)}, pk_{(\mathcal{U},ctx)}, pk'_{(\mathcal{U},ctx)})$ will output $pk_\mathcal{U}$. Then, $cert_\mathcal{U}$ can be revoked using $Revoke$. Note that we do not view a user who reuses the same public key $pk_{(\mathcal{U},ctx)}$ as a Sybil attacker. She is just using the same short term identity again.

5.2   Efficiency

The overall costs of our system are linear in the size of the identity domain with respect to users joining the domain, and quadratic with respect to the verification of the Sybil property: every user needs to execute the $Sign$ algorithm for herself, and the $Verify$ algorithm for all other users. The construction in [9] requires

10 multi-base exponentiations for pseudonym certificate creation and a similar number of multi-exponentiations for verification. Using multi-base exponentiation tricks, multi-base exponentiations can be made almost as efficient as normal exponentiations. This compares to schemes that do not support identification with about half the number of multi-exponentiations, and ordinary CA-issued pseudonym certificates with one or two exponentiations. Verification may not be needed in all cases, e. g., if users trust the domain controller to verify users on their behalf, or if the application bases its security properties on the assumption that only a set of key users are not Sybil nodes, rather than every single user.

5.3 Security Analysis

This section discusses some security properties of our proposed self-certified pseudonyms.

- *Sybil-Proof Property:* the cryptographic properties of e-token signatures ensure that for each valid membership certificate there can exist only one unique pseudonym $P_{(\mathcal{U},ctx)}$ per identity domain (see Section 4). However, as there is no inherent trust in any user in the identity domain (including the domain controller), users have to check the correctness of the pseudonym certificate $cert_{(\mathcal{U},ctx)}$ of all other users in the domain by locally running *Verify*. After an honest user has finished this verification and has checked the uniqueness of $P_{(\mathcal{U},ctx)}$, she is assured that her communication partner is a real user with public key $pk_{(\mathcal{U},ctx)}$, provided that she authenticated with $sk_{(\mathcal{U},ctx)}$.

- *Unlinkability Property:* our approach has strong unlinkability properties as the cryptographic properties of the e-token signatures ensure the algorithmic unlinkability of two pseudonym certificates generated for different domains (see Section 4). However, should the users violate precautions on the network or application layers, the attacker may still be able to make an educated guess on whether two arbitrary pseudonym certificates from different identity domains are related or not. In a real word scenario, a variety of different information could help the attacker to make such a guess, for instance, the location property of the identity domain or the location of the user. A traffic analysis of each setting is required to assess the concrete threats to the users' privacy.

6 Summary and Outlook

In this paper, we have described the construction of a solution to the Sybil attack that does not require online connectivity to a TTP but preserves user privacy: self-certified Sybil-free pseudonyms. We have discussed some real-world applications that would benefit from our solution. Future work includes implementing the proposed solution in a real system.

Acknowledgments

This research was funded by the European Network of Excellence Future of Identity in the Information Society (FIDIS) and by the European Integrated Project for Privacy and Identity Management for Europe (PRIME), both under the 6th Framework Program for Research and Technological Development within the Information Society Technologies (IST) priority.

References

1. Andersson, C., Camenisch, J., Crane, S., Fischer-Hübner, S., Leenes, R., Pearson, S., Pettersson, J.S., Sommer, D.: Trust in PRIME. In: Proceedings of the Fifth IEEE International Symposium on Signal Processing and Information Technology, pp. 552–559 (2005)
2. Anonymous Authors. Anonymized title. In: Proceedings of Anonymized Conference (2008)
3. Bangerter, E., Camenisch, J., Maurer, U.M.: Efficient Proofs of Knowledge of Discrete Logarithms and Representations in Groups with Hidden Order. In: Vaudenay, S. (ed.) PKC 2005. LNCS, vol. 3386, pp. 154–171. Springer, Heidelberg (2005)
4. Bhargav-Spantzel, A., Camenisch, J., Gross, T., Sommer, D.: User centricity: a taxonomy and open issues. In: DIM 2006: Proceedings of the second ACM workshop on Digital identity management, pp. 1–10. ACM Press, New York (2006)
5. Borcea-Pfitzmann, K., Franz, E., Pfitzmann, A.: Usable presentation of secure pseudonyms. In: DIM 2005: Proceedings of the 2005 workshop on Digital identity management, pp. 70–76. ACM Press, New York (2005)
6. Boudot, F.: Efficient Proofs that a Committed Number Lies in an Interval. In: Preneel, B. (ed.) EUROCRYPT 2000. LNCS, vol. 1807, pp. 431–444. Springer, Heidelberg (2000)
7. Brands, S.: Rapid Demonstration of Linear Relations Connected by Boolean Operators. In: Fumy, W. (ed.) EUROCRYPT 1997. LNCS, vol. 1233, pp. 318–333. Springer, Heidelberg (1997)
8. Brickell, E., Camenisch, J., Chen, L.: Direct anonymous attestation. In: CCS 2004: Proceedings of the 11th ACM conference on Computer and communications security, pp. 132–145. ACM Press, New York (2004)
9. Camenisch, J., Hohenberger, S., Kohlweiss, M., Lysyanskaya, A., Meyerovich, M.: How to win the clone wars: Efficient periodic n-times anonymous authentication. In: ACM Conference on Computer and Communications Security, ACM, New York (2006)
10. Camenisch, J., Hohenberger, S., Lysyanskaya, A.: Compact E-Cash. In: Cramer, R. (ed.) EUROCRYPT 2005. LNCS, vol. 3494, pp. 302–321. Springer, Heidelberg (2005)
11. Camenisch, J., Lysyanskaya, A.: Dynamic accumulators and application to efficient revocation of anonymous credentials. In: Yung, M. (ed.) CRYPTO 2002. LNCS, vol. 2442, pp. 61–76. Springer, Heidelberg (2002)
12. Camenisch, J., Lysyanskaya, A.: A signature scheme with efficient protocols. In: Cimato, S., Galdi, C., Persiano, G. (eds.) SCN 2002. LNCS, vol. 2576, pp. 268–289. Springer, Heidelberg (2003)

13. Camenisch, J., Lysyanskaya, A.: Signature schemes and anonymous credentials from bilinear maps. In: Franklin, M. (ed.) CRYPTO 2004. LNCS, vol. 3152, pp. 56–72. Springer, Heidelberg (2004)
14. Camenisch, J., Stadler, M.: Proof systems for general statements about discrete logarithms. Technical Report TR 260, Institute for Theoretical Computer Science, ETH Zürich (March 1997)
15. Damgård, I., Dupont, K., Pedersen, M.Ø.: Unclonable group identification. In: Vaudenay, S. (ed.) EUROCRYPT 2006. LNCS, vol. 4004, pp. 555–572. Springer, Heidelberg (2006)
16. Dodis, Y., Yampolskiy, A.: A Verifiable Random Function with Short Proofs and Keys. In: Vaudenay, S. (ed.) PKC 2005. LNCS, vol. 3386, pp. 416–431. Springer, Heidelberg (2005)
17. Douceur, J.R.: The Sybil Attack. In: Druschel, P., Kaashoek, M.F., Rowstron, A. (eds.) IPTPS 2002. LNCS, vol. 2429, pp. 251–260. Springer, Heidelberg (2002)
18. Fiat, A., Shamir, A.: How to Prove Yourself: Practical Solutions to Identification and Signature Problems. In: Odlyzko, A.M. (ed.) CRYPTO 1986. LNCS, vol. 263, pp. 186–194. Springer, Heidelberg (1987)
19. Franz, E., Borcea-Pfitzmann, K.: Intra-application partitioning in an elearning environment - a discussion of critical aspects. In: ARES '06: Proceedings of the First International Conference on Availability, Reliability and Security (ARES 2006), Washington, DC, USA, pp. 872–878. IEEE Computer Society, Los Alamitos (2006)
20. Fujisaki, E., Okamoto, T.: Statistical zero knowledge protocols to prove modular polynomial relations. In: Kaliski Jr., B.S. (ed.) CRYPTO 1997. LNCS, vol. 1294, pp. 16–30. Springer, Heidelberg (1997)
21. Kim, Y., Mazzocchi, D., Tsudik, G.: Admission control in peer groups. In: NCA, pp. 131–139. IEEE Computer Society, Los Alamitos (2003)
22. Kunz-Jacques, S., Martinet, G., Poupard, G., Stern, J.: Cryptanalysis of an efficient proof of knowledge of discrete logarithm. In: Yung, M., Dodis, Y., Kiayias, A., Malkin, T.G. (eds.) PKC 2006. LNCS, vol. 3958, pp. 27–43. Springer, Heidelberg (2006)
23. Levine, B.N., Shields, C., Margolin, N.B.: A survey of solutions to the sybil attack. Tech report 2006-052, University of Massachusetts Amherst, Amherst, MA (October 2006)
24. Pedersen, T.P.: Non-interactive and Information-Theoretic Secure Verifiable Secret Sharing. In: Feigenbaum, J. (ed.) Crypto 1992. LNCS, vol. 576, pp. 129–140. Springer, Heidelberg (1992)
25. Saxena, N., Tsudik, G., Yi, J.H.: Admission control in peer-to-peer: design and performance evaluation. In: Setia, S., Swarup, V. (eds.) SASN, pp. 104–113. ACM, New York (2003)
26. Saxena, N., Tsudik, G., Yi, J.H.: Efficient node admission for short-lived mobile ad hoc networks. In: ICNP, pp. 269–278. IEEE Computer Society, Los Alamitos (2005)
27. Schnorr, C.P.: Efficient signature generation for smart cards. Journal of Cryptology 4(3), 239–252 (1991)
28. Teranishi, I., Furukawa, J., Sako, K.: k-times anonymous authentication (extended abstract). In: Lee, P.J. (ed.) ASIACRYPT 2004. LNCS, vol. 3329, pp. 308–322. Springer, Heidelberg (2004)
29. Yu, H., Kaminsky, M., Gibbons, P.B., Flaxman, A.: SybilGuard: defending against sybil attacks via social networks. In: SIGCOMM 2006, pp. 267–278. ACM Press, New York (2006)

A   Cryptographic Building Blocks

A *zero-knowledge* (ZK) proof is an interactive proof in which the verifier learns nothing besides the fact that the statement that is proven is true. This notion is defined by means of a *simulator*, which can reproduce the communication knowing only what the verifier knows. A *proof of knowledge* is an interactive proof in which the prover succeeds in *convincing* a verifier that it knows something. What it means for a machine to *know something* is defined in terms of computation. A machine *knows something*, if this something can be computed, given the machine as an input. The machine extracting the knowledge is called the *knowledge extractor*. Protocols with a simulator and a knowledge extractor are called zero-knowledge proofs of knowledge.

For some protocols only simulators that work for honest verifiers are known. These are verifiers that choose the challenge according to a predetermined distribution. Honest-verifier zero-knowledge proofs-of-knowledge protocols that have a three move structure – commitment, challenge and response – are called *sigma protocols*. Such protocols can be made non-interactive by applying a cryptographic trick called *Fiat-Shamir heuristic* [18]. This heuristic uses a cryptographic hash function to allow the prover to compute the challenge herself without involving the verifier. Non-interactive proofs of knowledge have the advantage that they do not require interaction between the prover and the verifier. In addition, they allow to sign any message by hashing it together with the first message when creating the challenge.

Sigma protocols exist for proving knowledge of discrete logarithm (DL), equality of DLs, and linear relations between DLs in groups of known [27,7,14], and hidden order [3,22]. This allows us to prove statements about certain algorithms (some of wich are detailed below) that operate in these groups, for instance that two commitments contain the same value or that a committed value lies in a certain interval [6], that we know a signature for a value or a committed value, that a value was verifiable encrypted, or that a value was correctly created using a pseudo-random function and a secret seed.

B   Cryptographic Primitives

DY Pseudorandom Function (PRF). Let $\mathbb{G} = \langle g \rangle$ be a group of prime order $q \in \Theta(2^k)$. Let a be a random element of $\mathbb{Z}_q^*$. Dodis and Yampolskiy [16] showed that $f_{g,a}^{DY}(x) = g^{1/(a+x)}$ is a pseudorandom function, under the decisional Diffie-Hellman inversion assumption (y-DDHI), when either: (1) the inputs are drawn from the restricted domain $\{0,1\}^{O(\log k)}$ only, or (2) the adversary specifies a polynomial-sized set of inputs from $\mathbb{Z}_q^*$ *before* a function is selected from the PRF family (i.e., before the value a is selected). For our purposes, we require something stronger: that the DY construction work for inputs drawn arbitrarily and adaptively from $\mathbb{Z}_q^*$. Dodis-Yampolskiy PRF is adaptively secure for inputs in $\mathbb{Z}_q^*$ under the SDDHI assumption [9].

Pedersen and Fujisaki-Okamoto Commitments. Recall the Pedersen commitment scheme [24], in which the public parameters are a group $\mathbb{G}$ of prime order q, and generators $(g_0, \ldots, g_m)$. To commit to the values $(v_1, \ldots, v_m) \in \mathbb{Z}_q{}^m$, pick a random $r \in \mathbb{Z}_q$ and set $C = PedCom(v_1, \ldots, v_m; r) = g_0^r \prod_{i=1}^m g_i^{v_i}$. Fujisaki and Okamoto [20] showed how to expand this scheme to composite order groups.

CL Signatures. The Camenisch and Lysyanskaya signature scheme [12] includes two protocols: (1) An efficient protocol for a user to obtain a signature on the value in a Pedersen (or Fujisaki-Okamoto) commitment [24, 20] without the signer learning anything about the message. (2) An efficient proof of knowledge of a signature protocol. Security is based on the Strong RSA assumption. Using bilinear maps, we can use other signature schemes [13] for shorter signatures.

Availability for DHT-Based Overlay Networks with Unidirectional Routing

Jan Seedorf[1] and Christian Muus[2]

[1] NEC Laboratories Europe
Kurfuerstenanlage 36, 69115 Heidelberg, Germany
jan.seedorf@nw.neclab.eu
[2] University of Hamburg
Vogt-Koelln-Strasse 30, 22527 Hamburg, Germany
christian@muus.de

Abstract. *Distributed Hash Tables (DHTs)* provide a formally defined structure for overlay networks to store and retrieve content. However, handling malicious nodes which intentionally disrupt the DHT's functionality is still a research challenge. One particular problem - which is the scope of this paper - is providing availability of the DHT's lookup service in the presence of attackers. We focus on DHTs with unidirectional routing and present concrete algorithms to extend one particular such DHT, namely Chord. Our extensions provide independent multipath routing and enable routing to replica roots despite attackers on the regular routing path. In addition, we investigate algorithms to detect adversary nodes which employ node-ID suppression attacks during routing. We demonstrate how these techniques can be combined to increase lookup success in a network under attack by deriving analytical bounds for our proposed extensions and simulating how our algorithms come close to these bounds.

1 Introduction

Distributed Hash Tables (DHTs) [12] [13] [19] [20] offer a formally defined substrate for structured overlay networks to efficiently and consistently store data items. However, in general it cannot be guaranteed that nodes in the network behave according to the DHT-protocol. This opens the door for a broad range of attacks on DHTs [2] [7] [16].

Our contribution is the enhancement of a DHT with unidirectional routing so that it can handle a high degree of adversary nodes in the network and still provide successful lookups. Unidirectional routing has the advantage that all routing paths for a particular resource converge towards the node in the network responsible for storing that resource. While this is a disadvantage from a security perspective (as we will show) this property is beneficial for caching frequently queried resources [9]. We present concrete algorithms to extend a particular DHT, namely Chord [19], while preserving an unidirectional routing structure. Furthermore, we provide a theoretical analysis of our solutions and exhibit simulation results to show the effectiveness of our algorithms.

J.A. Onieva et al. (Eds.): WISTP 2008, LNCS 5019, pp. 78–91, 2008.

In Section 2 we discuss related work and compare it to our approach. Section 3 presents a formal DHT model and our attacker model. In section 4 we define the scope of our work: lookup availability. Section 5 presents Chord, the rationale why we chose this DHT, and theoretical results on lookup availability in Chord. We present concrete algorithms for Chord extensions in section 6, including simulation results. Section 7 concludes the paper with a short summary.

2   Related Work

Much work on various DHT security challenges exists. Here we survey previous work with focus on DHT lookup availability (the scope of our work).

Srivatsa and Liu present an analytical model for the failure rate of an arbitrary lookup in DHTs [18]. They derive theoretical bounds but do not provide concrete algorithms. In a previous publication we showed that for unidirectional DHTs stronger bounds can be obtained [15].

Castro et al. investigate lookup availability in a multidirectional DHT (Pastry [13]) [2]. They suggest constrained routing tables against routing table poisoning. Further, they rely on multipath routing to derive techniques for recursive routing in a multidirectional DHT which explore alternate routing paths. In contrary, we investigate an unidirectional DHT (Chord [19]) which does not provide multipath routing. Therefore, our problem domain is different and some of our solutions are specific to unidirectional DHTs.

Danezis et al. use a weak form of a social network, the bootstrap graph, to improve lookup performance in a Chord network under attack [6]. Marti et al. use an external, existing social network to increase the lookup success rate in Chord [10]. Hence, unlike our approach, the approaches in [6] and [10] rely on the existence of a social network to increase lookup availability. However, both of these approaches are complementary to our approach and we consider using these techniques as add-ons to our algorithms interesting future research.

The approach closest to ours is Cyclone [14], an extension to Chord which can guarantee multiple independent paths in Chord in the special case where the ID-space is fully utilised. Compared to Cyclone, our solutions are beneficial in any network, independent of ID-space utilisation. In addition, our work differs from the one in [14] because we use iterative routing (which allows the detection of node-ID suppression attacks) and we directly route to replicated content for increased availability.

Contrary to our work, none of the previous extensions to Chord [6] [10] [14] considers nor mitigates the case where the node responsible for storing content (or its predecessor) is an adversary node. Not only do we consider this case in our model, additionally we provide techniques to alleviate this problem. Further, our approach is the first to enable the detection of node-ID suppression attacks on every routing hop in Chord. For our extensions to Chord we assume that secure node-ID assignment against *Sybil attacks* [7] is used. Techniques for secure node-ID assignment in a DHT have been suggested by Awerbuch and Scheideler [1], Condie et al. [3], or Fiat et al. [8] and are outside the scope of this paper.

3   Formal DHT Model and Attacker Model

The two basic primitives provided by a DHT are *store (key,data)*, and *lookup (key) = data*. DHTs have been designed to guarantee consistent data storage and load balancing even when nodes enter and leave the network at a high frequency. Examples of Distributed Hash Tables are CAN [12], Pastry [13], Chord [19], and Tapestry [20]. To be able to classify threats on DHTs, precisely define the scope of our work, and to formally specify our extensions to *Chord* we use a formal model of a DHT.

3.1   Formal DHT Model

Our formal model of a DHT consists of the following:

Node-ID and Key-ID Space: An $l-bit$ *key identifier space* K and an $m-bit$ *node identifier space* I define the basic DHT structure. The DHT provides a function for mapping a key onto a key-ID k, $f_{km}(key) = k \in K$ and rules for mapping an external identifier eID onto a node identifier *(node-ID)* n_i, $f_{nm}(eID) = n_i \in I$.

Data responsibility: A data placement function $f_{dp} : K \to I$ maps a key-ID $k \in K$ onto the node-ID space I and a responsibility function $f_{resp} : I \to I$ states which node $n_i \in I$ is responsible for storing $f_{dp}(k)$. Thus, the data item for key k is stored at node $n_i = f_{resp}(f_{dp}(k))$. We denote the node responsible for storing data belonging to key k as the root node for that key $root_k$. For reliability, a replication function $f_{rep} : I \to I^r$ maps the key onto r other nodes which store the data for k as well; we call these nodes the replica roots for key k denoted by $root_k \ldots root_k^r$.

Routing: A routing table T_r at each node n_i contains t links to nodes at some distance in the ID-space. Further, a second routing table T_s at each node n_i contains s direct neighbors in the DHT structure. Routing table functions $f_{tr} : I \to I^t$ and $f_{ts} : I \to I^s$ determine which nodes are in T_r and T_s of any node n_i in the system. A routing function $f_{route} : K \to I$ specifies which entry the routing table returns upon receiving a message (lookup or storage) for key-ID k.

State: Since the system is dynamic, its state changes constantly and a set of rules for joining and leaving of nodes is necessary. As we do not examine joining and leaving of nodes in this paper we do not define these rules formally. Σ denotes the set of possible states. At any state $\sigma_i \in \Sigma$ we have N nodes in the system. The set of all N nodes (denoted $N \subseteq I$), their N routing tables T_r and T_s, and all the data items stored in the system define the current state σ_i.

3.2   Attacker Model

We assume the following attacker model: A network consisting of only good nodes is infiltrated over a certain period of time by attacker nodes which either join the system or compromise good nodes. After this period, at a certain state $\sigma_i \in \Sigma$ the system contains $N_a = f*N$ adversary nodes and $N_g = (1 - f)*N$ good nodes, where $f < 1$ and $N_a \cap N_g = \oslash$. All adversary nodes may collude (e.g., because they are controlled by a single external identity). Adversary nodes route exclusively to adversary nodes and do not drop messages: $\forall n_i \in N_a :$ $f_{route}(k) = n_j \in N_a$ (i.e., adversary node *suppress* existing good nodes in their routing tables); good nodes route to good and adversary nodes: $\forall n_i \in N_g :$ $f_{route}(k) = n_j \in N$.

In principle, adversary nodes could also drop messages. However, this would result in a less severe attack on lookup availability because this behaviour can easily be detected through time-outs. In contrary, by continuing to route amongst them (never reaching the target data item) colluding adversary nodes can absorb more DHT routing resources in vain. Thus, by expecting adversary nodes not to drop messages we consider a stronger attacker model.

Adversary nodes are distributed uniformly[1] over the node-ID space I. Additionally, we assume that any message sent on a single DHT-hop will arrive unchanged (i.e., attacks on the IP-layer are out of scope).

4   Availability of the Lookup Service

In principle, without a trusted authority in the network, a single adversary can control a large fraction of an overlay network with only a few external identities [7]. An adversary node on the path from the query node to some key can either drop the message, alter the message, or route the message to another adversary node. Castro et al. were the first to thoroughly investigate this problem [2]. They conclude that in order to achieve *secure routing* in a DHT three properties have to be fulfilled: 1) *Secure node-ID assignment*, 2) *Protection against routing table poisoning*, and 3) *Secure message forwarding*. In this context we define *lookup availability* as follows:

Definition 1 *The* Availability of the Lookup Service *is the probability that the corresponding data item is returned by the DHT after a node has invoked an arbitrary lookup for a key.*

A lookup can consist of many routing attempts from the query node to the key. Thus, a lookup can use several different paths and is finished if either it succeeds, a threshold t_h (limiting the number of hops used in the lookup) is reached, or all possible paths between the query node and the node responsible for storing the corresponding data item (i.e., $root_k$) have been tried without success. We define a path in a DHT as follows:

---

[1] Existing work on secure node-ID assignment for DHTs and for Chord in particular [1] [2] [3] [8] provides solutions to achieve this property. Thus, this assumption is reasonable if secure node-ID assignment techniques are used. We expect the use of such techniques as a fundament for our extensions (see further section 4).

Definition 2 *A path $p(n_q, k) \subseteq N$ from a query node $n_q \in N$ for key $k \in K$ is any set of nodes such that routing from n_q for key k will pass through these nodes including $root_k$. Two different paths are called* alternate *if at least one node (other than n_q and $root_k$) is on both these paths and* independent *if they share no common node other than n_q and $root_k$.*

As a metric for lookup availability in a DHT we use the success-rate of a random lookup (as a secondary metric we use the hop count of a random lookup, denoted with χ):

Definition 3 *The* success rate ρ *is the probability that a random lookup will succeed:* $\rho = P(\exists p(n_q, k) | \forall n_i \in p(n_q, k) : n_i$ *is good) where n_q is a random query node and k is a random key.*

We assume that secure Node-ID assignment techniques against *Sybil attacks* [7] are used [1] [2] [3] [8] and that the DHT is protected against routing table poisoning (*Eclipse attacks* [16]): $f_{nm}()$, $f_{tr}()$, and $f_{ts}()$ cannot be attacked. This implies that at state σ_i in a reasonably large network the routing table T_r of any good node in the system contains with high probability $f \times d$ adversary nodes and $(1 - f) \times d$ good nodes, where $d \leq t$ is the number of distinctive nodes in T_r. Further, we assume that the integrity of data items stored in the DHT can be verified by the application on top of the DHT, e.g., by using a public key infrastructure or self-certifying keys/data [5].

Despite these assumptions attackers are still able to degrade the availability of the DHT severely by attacking the routing function $f_{route}()$, i.e., message forwarding. Our goal is to develop algorithms for $f_{route}()$ that provide resilience against such attacks on the DHT-layer.

5   Extending an Existing Unidirectional DHT

As an example DHT with unidirectional routing we choose Chord [19]. Our goal is to make as few general changes to regular Chord as necessary. In fact, we only make very few changes to Chord that have to be adopted by all nodes in the system (which we call *global* extensions). These changes do not change Chord's formal properties. Most extensions we introduce are *local*: nodes can optionally decide to use a different $f_{route}()$ function than in regular Chord. However, these local extensions do not affect other nodes or the DHT.

Chord uses the IP-address of a node as its external identifier (*eID*). A pre-defined hash function $h()$ maps any *eID* onto an $m - bit$ node-ID n_i and also any key onto an $m - bit$ key-ID k. The node identifier space I is a virtual ring where node-IDs are ordered clockwise from 0 to $2^m - 1$. Each node in the ring is responsible for storing the content of all key-IDs that are equal to or less than its own identifier but larger than the identifier of the node's direct predecessor in the Chord ring. For reliability against node failures, the data for k is also stored at r nodes directly succeeding $root_k$ in the ring. In its routing table T_r each

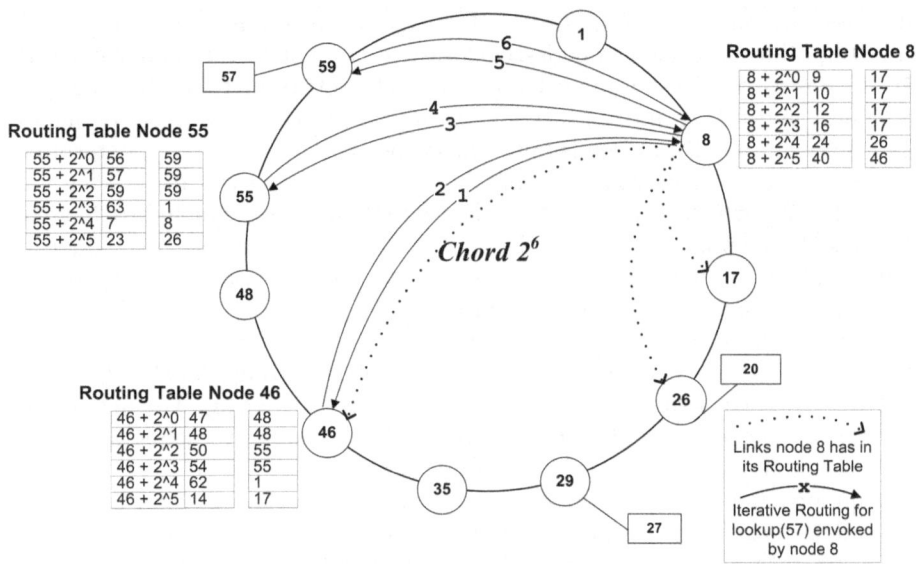

Fig. 1. Iterative Routing in Chord

node n_i stores links to m succeeding nodes in the ring (*unidirectional routing* [9]). Additionally, each node keeps a link to its direct predecessor in the ring.

It is precisely specified how routing tables are filled (making routing tables *constrained*, protecting against *Eclipse attacks* [2] [16] and thus making our assumption of $f_{tr}()$ and $f_{ts}()$ being secure reasonable for Chord): The jth entry in T_r contains the IP-address of the first node that follows n_i by at least 2^{j-1} in the virtual ring: $f_{tr}(n_i) = [succ(n_i + 2^0), succ(n_i + 2^1), ..., succ(n_i + 2^{m-1})]$ where $succ(x) = n_o \geq x(\neg \exists n_p \in N | n_o > n_p \geq x)$. '$>$' is a relation 'succeeding in the ring' using modular arithmetic to ensure that routing and data responsibility is shared across $2^m - 1$ in the ring. The first entry in T_r is the node directly succeeding n_i. The last entry in T_r contains a link to a node at least $\frac{2^m}{2}$ away from n_i in the ring. To achieve fast lookups, nodes forward messages to the node with the highest ID in their routing table that is smaller than the key-ID (greedy routing). Routing succeeds when the direct successor of a node has a larger ID than the key-ID. This successor node is responsible for the key. Additionally, each node keeps a list of its s direct successors T_s to handle node failures. Routing is either iterative (the query node contacts other nodes to get iteratively closer to the key) or recursive (a query message is passed through the network hop by hop).

Figure 1 exemplifies iterative routing in a Chord network [15]. In the routing tables displayed the rightmost column shows to which other nodes in the DHT links exist. The two leftmost columns point out how to compute precisely which node is in the particular routing table entry, i.e., determining the value where the first node 'succeeding' this value in the ring must be in that routing table entry (compare to the previous paragraph). In this example, a query node with node-ID 8, n_8, starts a lookup for key-ID 57. n_8 sends a message to the node in its routing table that has the node-ID closest to but not larger than the key-ID (57) which is n_{46} in this example $\langle 1 \rangle$. n_{46} replies by returning to n_8 the node with the highest node-ID from its routing table not larger than the key-ID which is n_{55} in this example $\langle 2 \rangle$. n_8 sends out a query message to n_{55} $\langle 3 \rangle$. n_{55} determines that the first node in its routing table, i.e., its direct successor in the ring, has a node-ID (59) which is higher than the key-ID (57). n_{55} concludes that this node must be responsible for key-ID 57 and returns n_{59} to the query node n_8 $\langle 4 \rangle$. To retrieve the data item for key-ID 57, n_8 contacts n_{59} $\langle 5 \rangle$ which answers by sending the corresponding data item to n_8 $\langle 6 \rangle$.

In regular Chord, any lookup has to pass the predecessor of the node storing the content for the key looked up. This is also referred to as the *shield problem* [11] [15] and a consequence of unidirectional greedy routing. We denote the predecessor of $root_k$ with $shield_k$ for any key k. Formally, we define $shield_k = n_i \in N | (n_i < root_k) \wedge (\neg \exists n_l \in N | n_i < n_l < root_k)$.

An important consequence of the shield problem is that in Chord only one independent path from the query node n_q to $root_k$ exists for any lookup. Hence, the success rate for an arbitrary lookup in regular Chord is bound by the following inequality [15]:

$$P(lookupsuccess) \le (1 - f)^2 \tag{1}$$

To see why inequality (1) holds, consider a random lookup for a key. In our model, with probability $(1 - f)$, $root_k$ is good and with probability $(1 - f)$, $shield_k$ is good. Any lookup can only succeed if both nodes are good because any lookup has to pass these two specific nodes. Since it is statistically independent if either one of the two nodes is controlled by an adversary equation (1) holds.

In [18] an upper bound for DHTs is given on the failure rate for an arbitrary lookup which can be converted into a lower bound on the success rate by taking the opposite event and mapped to Chord [15]:

$$1 - \left(1 - (1 - f)^{\left(\frac{1}{2}\right) \log N}\right) \le P(lookupsuccess) \tag{2}$$

6   Algorithms for Increased Lookup Availability

In this section we describe our extensions to Chord for increasing lookup availability. In principle, we combine three techniques: 1) We use the direct successor list of each node to accomplish independent multipath routing. 2) To overcome the shield problem we directly route to replica roots. 3) We use density checks on each iterative routing hop to detect paths that contain adversary nodes as early as possible.

For all our techniques described below we use the following general (global) extension to Chord: Each node in the network must support iterative routing where at each routing hop the query node receives not only the next hop from the node it queried (as in regular Chord) but instead the whole routing table T_r of the queried node and its list of direct successors T_s. Note that this extension only affects the size of each iterative query response. In particular, it does not affect the total number of links stored at each node because the additional information received at each iterative routing step is only stored temporarily during the lookup. We call this extension *complete-knowledge iterative routing* because at each iterative routing hop the query node receives the complete information the hop node has about the network. All other routing techniques we introduce are solely computed at the query node (locally). Thus, it does not affect the success rate of a lookup if other nodes in the network use these techniques or not.

6.1   Multiple Independent Paths

In the case a lookup path has failed, we explore two techniques to let the lookup continue (we refer to this as *failover routing*)[2]: a) by starting a new independent path at the query node (*independent restart*) or b) by starting a new path at the closest node to the key received during the previous path which has not been used in the lookup (*backtracking*).

For both techniques the query node maintains a temporary list T_m of nodes it has used in the lookup so far. In each individual path it explores during a lookup the query node only uses nodes it has not used before in this lookup, i.e., nodes $\notin T_m$. In regular Chord the direct successor list T_s is only used for redundancy (i.e., in the case of node failures). We allow each node to use the list of direct successors T_s on every routing hop. Since we use complete-knowledge iterative routing a query node can in principle use for the next hop any node from the routing table T_r and the direct successor list T_s it received from the node on the last hop. However, for our extensions at each hop the nodes in T_s are only used in routing if all nodes from T_r have been used previously in the lookup, i.e., are already in T_m. n_q (the query node) always routes greedy (as in regular Chord): It always uses the node $n_i \in T_r$ (or $n_i \in T_s$ if $\forall n_i \in T_r : n_i \in T_m$) with the highest node-ID smaller than k. This assures that queries make progress.

With unidirectional greedy routing independent paths converge towards the root [9]. Using T_s allows a path to continue if at some hop in T_r all entries smaller than k are already in T_m. For independent restart, using T_m guarantees that all paths in a lookup are independent. Further, independent restart allows for up to s (the number of entries in every T_s) independent paths because this is the maximum number of independent paths that can converge on the penultimate hop before reaching the root. Because with backtracking a new path does not start at the query node, this technique explores alternate (not independent) paths.

In our model, adversary nodes suppress good nodes in the routing tables T_r and T_s they return. This implies that once a path has reached an adversary node, only adversary nodes will be added to T_m on this path. Thus, node-ID suppression attacks do not prevent our technique to subsequently explore a path with only good nodes on every hop.

---

[2] Remember that in our model a lookup consists of several individual paths.

6.2   Direct Replica Routing

To tackle the situation where $root_k$, $shield_k$, or both are malicious we allow to route directly to the replica roots of k. Chord replicates content at r replica roots which are the r nodes directly succeeding $root_k$ in the ring. However, in regular Chord the replica roots are only used for redundancy (i.e., node failure of $root_k$).

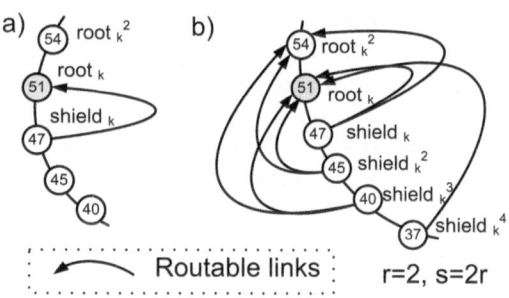

Fig. 2. Direct Replica Routing

We now extend Chord in a way that routing to the replica roots of a key k is possible without passing $shield_k$ nor $root_k$: We allow *direct* routing to a node $n_i \in root_k \ldots root_k^r = REP_k$ if $n_i \in T_s$ (we refer to this as *direct replica routing*). Because at every hop T_s contains s direct successors in the ring, the query node can check if some of these nodes are $\in REP_k$ (n_q simply has to verify if $\exists n_j \in T_s | k \leq n_j \leq root_k^r$). If all replica roots retrieved at some hop have been queried without success, a failover (backtracking or restart) is pursued.

Using direct replica routing results in each key k having effectively s shield nodes (the s direct predecessors of $root_k$) which we denote with $shield_k \ldots shield_k^s = SHI_k$. By setting $s = 2r$ (globally) in the system, any of the $r+1$ closest shield nodes to a particular key k can route directly to any of the r replica root nodes for k. In general, setting $s \geq r$ ensures that the last replica root $root_k^r$ is accessible from $s - r + 1$ shield nodes.

Figure 2 exemplifies how replica roots can be reached through more than one node (b) compared to regular Chord (a). Any T_s the query node n_q will receive from an adversary node will only contain the next s adversary nodes in the ring. However, by setting $s = 2r$ we guarantee that reaching one good shield node of the r closest shield nodes to k is enough to reach one good replica root $\in REP_k$ (if existing).

6.3   Detecting Node-ID Suppression Attacks

Recall that in our attacker model a network of good nodes is infiltrated and routing tables in Chord are constrained (and therefore protected against routing table poisoning). Thus, good nodes have (with high probability) $f \times d$ adversary nodes and $(1-f) \times d$ good nodes in their routing table T_r as well as $f \times s$ adversary nodes and $(1-f) \times s$ good nodes in T_s. Adversary nodes suppress good nodes in the routing tables they return. This enables them to attack lookup availability even if complete-knowledge iterative routing is used by the query node.

We can detect these attacks by using *density checks*: the query node n_q calculates the average distance α between nodes in its direct successor list T_s as

$\alpha = \frac{n_l - n_f}{s}$ where n_l is the last entry in T_s and n_f is the first entry in T_s. From any routing table $T_s(n_i)$ that n_q receives from a node n_i, n_q can compute $\alpha(n_i)$ and compare it with its own average distance by computing $\delta = \frac{\alpha(n_i)}{\alpha(n_q)}$. If $\delta \geq t_d$ (a density threshold), n_q considers n_i to be an adversary node.

An adversary node n_a can only decrease its $\alpha(n_a)$ by either creating artificial entries in $T_s(n_a)$ (which will be detected on the next hop if such an entry is chosen by n_q) or limit suppression of good nodes in $T_s(n_a)$ (which would give n_q access to good nodes). With a low density threshold t_d there is a risk of falsely estimating good nodes as adversary ones. However, this only affects $f_{route}()$ of n_q locally.

6.4   Theoretical Analysis of the Proposed Extensions

Our proposed extensions to Chord provide several independent paths between n_q and $root_k$, and route directly to the replica roots of a key k so that not a single root node can control all access to data items for a key k. Thus, there exist at most s shield nodes (one on the penultimate hop of every independent path) denoted $shield_k \ldots shield_k^s$ and for every key k there are r routable replica roots, denoted $root_k \ldots root_k^r$.

We now extend the theoretical results for regular Chord from Section 5 to this case. Analytically, we use a sample space Ω for a random lookup. Ω samples all shields and all replica roots for an arbitrary key and determines for each shield and replica root node if it is an adversary node. We are interested in the following events in our sample space:

$A = \{$"at least one shield node is good"$\}$ $B = \{$"at least one replica root is good"$\}$

$E = \{$"at least one replica root and one shield node are not adversary nodes"$\}$

Event E states an upper bound on the success rate for an arbitrary lookup because this event is the minimum requirement for any lookup to succeed (a lookup can still fail under this event if all paths explored contain at least one adversary node).

We now derive the probability for event E for the case that we have precisely s shield nodes and r routable replica roots for any key k:

$$P(A) = 1 - f^s \quad P(B) = 1 - f^r \tag{3}$$

$$P(lookup success) \leq P(E) = P(A) * P(B) = (1 - f^s) * (1 - f^r) \tag{4}$$

Note that it is possible to multiply P(A) and P(B) because these events are statistically independent in our model. Adapting the lower bound from inequality (2) to s independent paths we get [18]

$$1 - \left(1 - (1 - f)^{\left(\frac{1}{2}\right)\log N}\right)^s \leq P(lookup success) \tag{5}$$

With our extensions, there exist at most s independent paths and exactly r replica roots. Since equation (4) provides an upper bound, it holds for our extensions even though some lookups might explore less than s independent paths.

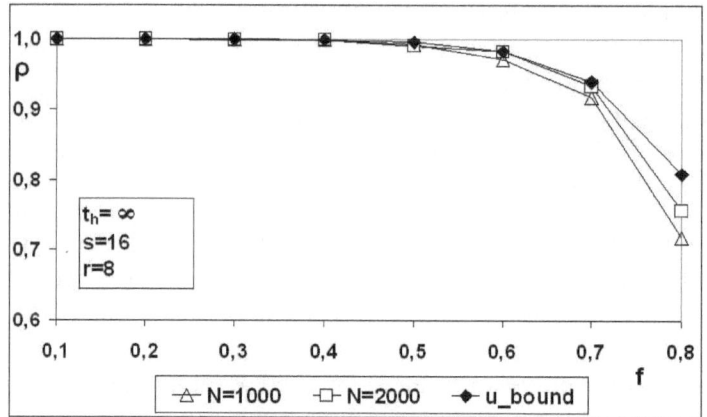

Fig. 3. $MRR - r$ with $t_h = \infty$ compared to theoretical upper bound (N= 1000/2000)

However, the lower bound in inequality (5) does not apply to our extensions. Still, it indicates analytically that as more independent paths are explored (which is the effect of our multipath-extensions to Chord) the lower bound on the success rate increases. In any case, we are interested in the maximum success rate (and thus the upper bound) that our extensions can theoretically achieve.

6.5   Simulations

To see how close our algorithms come to theoretical limits we simulated multipath routing combined with direct replica routing (which we call MRR for *Multipath Replica Routing*) for various network sizes N and attacker rates f and compared it to the upper bound on ρ from equation (4). In all our experiments we simulated 1000 lookups in 10 random Chord networks with $|I| = 2^{32}$ and adversary nodes behaving according to our attacker model. We only consider lookups where $T_s(n_q) \cap REP_k = \varnothing$, i.e., lookups where no replica root is contained in the direct successor list of the query node.

Figure 3 shows results for independent restart (we also conducted simulations for backtracking with very similar results). It can be observed that our algorithms come very close to the upper bound (u_bound) on lookup success in equation (4), almost reaching theoretical limits even for high attacker rates.

We noticed however that with $t_h = \infty$ (as in Figure 3) the average hop count χ can get quite high with increasing levels of network infiltration (e.g., for $f = 0.7$, N= 2000 and a success rate of 92% we obtained an average of 635 hops per lookup). In some applications for which DHTs have been proposed (e.g., signalling in real-time communications [17] or a distributed DNS architecture [4]) the time it takes for a lookup to succeed is crucial. To reflect this requirement and investigate the effectiveness of our algorithms with a timing constraint, we conducted simulations with a hop threshold t_h. Figure 4 displays MRR with backtracking (-b) and independent restart (-r) compared to regular Chord with

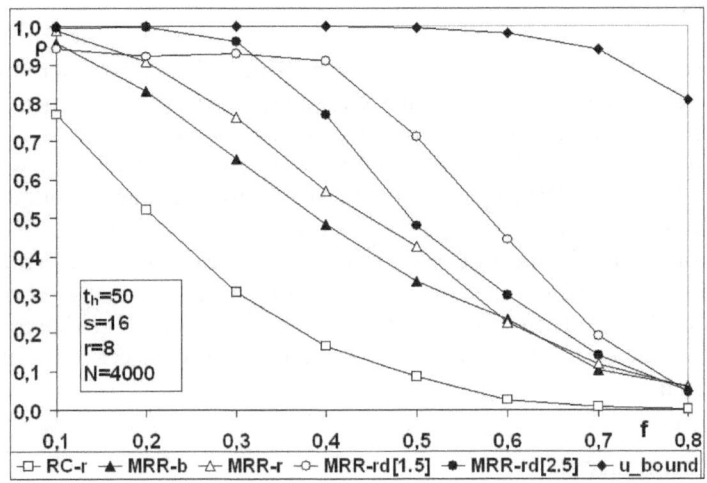

Fig. 4. Success rate for MRR compared to regular Chord and upper bound ($t_h = 50$, N= 4000)

independent restart (RC-r) for $t_h = 50$. Additionally, the figure shows the success rate for MRR-r with density checks (MRR-rd) for $t_d = [1.5, 2.5]$.

It can be noticed that independent restart performs better than backtracking for attacker rates up to $f = 0.6$. Further, the detection of node-ID suppression attacks with density checks on every hop increases lookup availability perceptibly. One can see that a higher threshold t_d is better suited for low attacker rates whereas a lower threshold results in better performance for high attacker rates (in Figure(4), for attacker rates higher than $f = 0.3$, $t_d = 1.5$ performs better). In general, it is advisable to set $t_d < \frac{1}{f}$ because the range of an attacker's successor list increases reciprocal to f with node-ID suppression attacks.

In addition to increasing the success rate, density checks also significantly decrease the hop count χ. Table 1 illustrates this by showing ρ and χ for MRR-r (with and without density checks) with f=0.6 and N=2000. Compared to MRR-r without any hop threshold, density checks achieve a more than 35 % lower success rate. However, note that the average hop count needed for this result is a factor of 5 lower. Compared to MRR-r using the same hop threshold ($t_h = $

Table 1. ρ and χ for MRR-r (f=0.6, N=2000)

	t_h	t_d	ρ	χ
MRR-r	100	∞	0.49	74.1
MRR-r	∞	∞	0.98	321
MRR-rd	100	1.5	0.62	59.8
MRR-rd	100	2.5	0.61	68.1

100), routing with density checks requires $\sim$14/6 less hops on average ($t_d = $ 1.5/2.5) even though it achieves a higher success rate. We consider exploring the tradeoff between ρ, χ and t_h interesting future research (in the end, deciding on this tradeoff depends on application constraints/demands).

7   Conclusion

We enhanced a DHT with unidirectional routing (Chord) to increase lookup availability. Our proposed algorithms enhance Chord with independent multipath routing, direct routing to replica roots, and mechanisms for detecting node-ID suppression attacks to provide resilience of the DHT's lookup service against attacks on the DHT-routing layer. We showed through simulations that our algorithms can come very close to theoretical limits. For example, we can achieve a lookup success rate of 98 % in a network with 60 % adversary nodes.

We consider combining our algorithms with techniques relying on social networks on top of a DHT (see related work) as well as exploring the tradeoff between the hop threshold, the average hop count, and the success rate interesting future research.

Acknowledgement

The authors would like to thank Rolf Winter for providing useful suggestions to an earlier version of this paper and the anonymous reviewers for providing valuable comments which improved the quality of the paper.

References

1. Awerbuch, B., Scheideler, C.: Towards Scalable and Robust Overlay Networks. In: Sixth International Workshop on Peer-to-Peer Systems, IPTPS 2007, Bellevue, WA, USA, February 26-27 (2007)
2. Castro, M., Druschel, P., Ganesh, A., Rowstron, A., Wallach, D.S.: Secure routing for structured peer-to-peer overlay networks. In: Proc. of the 5th Symposium on Operating Systems Design and Implementation, Boston, MA, December 2002, ACM Press, New York (2002)
3. Condie, T., Kacholia, V., Sankararaman, S., Maniatis, P., Hellerstein, J.M.: Maelstrom: Churn as Shelter, University of California at Berkeley Technical Report No. UCB/EECS-2005-11 (November 2005)
4. Cox, R., Muthitacharoen, A., Morris, R.: Serving DNS Using a Peer-to-Peer Lookup Service. In: Druschel, P., Kaashoek, M.F., Rowstron, A. (eds.) IPTPS 2002. LNCS, vol. 2429, Springer, Heidelberg (2002)
5. Dabek, F., Kaashoek, M.F., Karger, D., Morris, R., Stoica, I.: Wide-area cooperative storage with CFS. In: Proc. of SOSP 2001, Banff, Canada (2001)
6. Danezis, G., Lesniewski-Laas, C., Kaashoek, M.F., Anderson, R.: Sybil-Resistant DHT Routing. In: de Vimercati, S., Syverson, P.F., Gollmann, D. (eds.) ESORICS 2005. LNCS, vol. 3679, pp. 305–318. Springer, Heidelberg (2005)
7. Douceur, J.R.: The Sybil Attack. In: Druschel, P., Kaashoek, M.F., Rowstron, A. (eds.) IPTPS 2002. LNCS, vol. 2429, Springer, Heidelberg (2002)
8. Fiat, A., Saia, J., Young, M.: Making Chord Robust to Byzantine Attacks. In: Brodal, G.S., Leonardi, S. (eds.) ESA 2005. LNCS, vol. 3669, pp. 803–814. Springer, Heidelberg (2005)
9. Lua, E.K., Crowcroft, J., Pias, M., Sharma, R., Lim, S.: A Survey and Comparison of Peer-to-Peer Overlay Network Schemes. IEEE Communications Surveys and Tutorials 7(2), 72–93 (2005)

10. Marti, S., Ganesan, P., Garcia-Molina, H.: DHT Routing Using Social Links. 3rd Int. Workshop on Peer-to-Peer Systems (2004)
11. Muus, C.: Availability in DHT-based Structured Overlay Networks Considering Chord as an Example, Diploma Thesis, University of Hamburg, Germany (November 2007)
12. Ratnasamy, S., Francis, P., Handley, M., Karp, R., Shenker, S.: A Scalable Content-Addressable Network. In: Proc. of SIGCOMM 2001, San Diego, USA, August 27-31 (2001)
13. Rowstron, A., Druschel, P.: Pastry: Scalable, decentralized object location and routing for large-scale peer-to-peer systems. In: Proc. of the 18th IFIP/ACM International Conference on Distributed Systems Platforms, Heidelberg, Germany (November 2001)
14. Sanchez Artigas, M., Lopez, P.G., Skarmeta, A.F.G.: A Novel Methodology for Constructing Secure Multipath Overlays. IEEE Internet Computing 9(6), 50–57 (2005)
15. Seedorf, J., Muus, C.: Availability for Structured Overlay Networks: Considerations for Simulation and a new Bound on Lookup Success. In: 12th Nordic Workshop on Secure IT-Systems - NordSec 2007, Reykjavik, Iceland (October 2007)
16. Singh, A., Castro, M., Druschel, P., Rowstron, A.: Defending against eclipse attacks on overlay networks. In: Proc. of the ACM SIGOPS European Workshop (September 2004)
17. Singh, K., Schulzrinne, H.: Peer-to-Peer Internet Telephony using SIP. In: Proc. of the international workshop on Network and operating systems support for digital audio and video, Stevenson, Washington, USA, June 2005, pp. 63–68. ACM Press, New York (2005)
18. Srivatsa, M., Liu, L.: Vulnerabilities and Security Threats in Structured Overlay Networks: A Quantitative Analysis. In: Proc. of the 20th Annual Computer Security Applications Conference (ACSAC), Tucson, Arizona, December 6-10, 2004, pp. 251–261. IEEE CS Press, Los Alamitos (2004)
19. Stoica, I., Morris, R., Liben-Nowell, D., Karger, D.R., Kaashoek, M.F., Dabek, F., Balakrishnan, H.: Chord: A Scalable Peer-to-Peer Lookup Protocol for Internet Applications. In: IEEE/ACM Transactions on Networking, February 2003, vol. 11(1), IEEE Press, Los Alamitos (2003)
20. Zhao, B.Y., Huang, L., Stribling, J., Rhea, S.C., Joseph, A.D., Kubiatowicz, J.: Tapestry: A Resilient Global-Scale Overlay for Service Deployment. IEEE Journal on Selected Areas in Communications 22(1) (January 2004)

Network Smart Card Performing U(SIM) Functionalities in AAA Protocol Architectures

Joaquin Torres, Antonio Izquierdo, Mildrey Carbonell, and Jose M. Sierra

Carlos III University of Madrid, Spain
Computer Science Department
{joaquin.torres,antonio.izquierdo,mildrey.carbonell,jm.sierra}@uc3m.es

Abstract. This paper reviews the way in which the security protocols EAP-SIM/AKA are used in 3G/WLAN network interworking from the point of wiew of the U(SIM). As result, a new AAA protocol architecture is derived from the integration of a Network Smart Card, NSC, that implements U(SIM) functionalities within the scheme. The implementation in a testbed shows the robustness and feasibility of such an architecture.

Keywords: Network smart cards, authentication and authorization architectures, WLAN/3G interworking, secure protocols.

1 Introduction

The wide availability of wireless equipments of reduced size increases the demand for access points to the worldwide digital information and services. Although initially WLANs were conceived as an extension of corporative networks, nowadays their usage has been popularized in SOHO, campus and residential environments. The number of public hotspots is continuously proliferating, and this allows the information to be accessible in any time and any place.

The third generation mobile systems could be seen as a competitive solution, in terms of wide geographical area coverage and effective roamings. Moreover, depending on the scenario, issues such as reliability, throughput, value-added services (e.g. global localization) and contents (including multimedia services directly to your mobile phone) should be considered as advantages that this technology could offer. However, the expensive investment required by the 3G networks forces to the operators to look for more profitable and versatile solutions, and aiming to offer a wider variety of services for avoiding a leakage of subscribers.

WLAN's features allow to provide services with significant transmission rates in high demand zones and when the mobility is not a requirement. On other hand, 3G systems offer high mobility, wide coverage, well-established voice services but lower transmission rates, so they are more adequate for low/medium demand. Additionally, these systems posse a robust network and management infrastructure to deal with demands for security, billing and roaming requisites. Thus, as it is shown, wireless local area and 3G networks are complementary.

J.A. Onieva et al. (Eds.): WISTP 2008, LNCS 5019, pp. 92–105, 2008.

The wireless local area and 3G cellular networks interworking is a clear trend in the public access infrastructures (*PWLAN , Public Wireless LAN*) [1], which are progressively being deployed. It is considered as a significant step towards the fourth generation of all-IP wireless networks.

The combination of both technologies are allowing the development of services with high transmission rates (e.g. IP-based multimedia services, IMS) in mobile/roaming scenarios for an important number of profiled subscribers and preserving the quality of services. Beyond multimode terminals that provide both wireless interfaces (3G and WLAN) in order to access to each system, there exist integral solutions that provide transparent roaming between both technologies by the appropriate smart switching, with the goal of keeping initiated sessions.

In the 3G/WLAN integration, the subscriber must be authenticated before being her access to network services authorized. Thus user's multimode devices (e.g. laptops, smartphones, PDAs, etc.) require the appropriate personalized secure module. As in the stand-alone 3G systems, the chip card-based U(SIM) provides this functionality in PWLANs.

The important role of smart cards in this context is worth studying if one considers potential scenarios with the corresponding security functionalities. In Figure 1, an independent smart card with authentication purposes is isolated in the reference model.

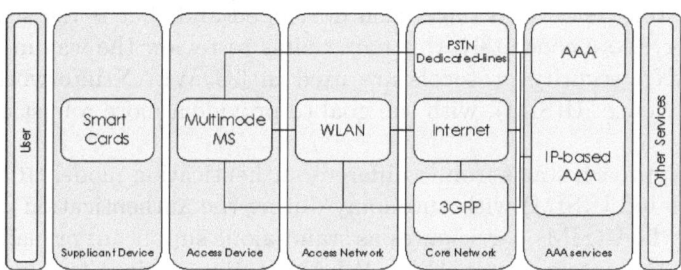

Fig. 1. Reference Model

In the 3G/WLAN interworking, the authentication schemes are based on a combination of the solutions that were initially supported by these two systems: the SIM-based solutions simultaneously inherit from EAPoL-based (i.e. 802.1X/EAP, RADIUS [2][3] or DIAMETER [4] used in WLAN technologies) and from U(SIM) authentication schemes supported by 3GPP subscriber registers (i.e. HLR/HSS).

The standardized protocols EAP-SIM [5] and EAP-AKA [6] represent the two most relevant SIM-based authentication schemes that establish mutual authentication between the mobile station and the backend authentication server. On one hand, the user is accustomed to use an (U)SIM, which allows her to access to a set of services by means of her mobile phone. On the other hand, the 3G/WLAN network operators do not require a different credential or secure module in order to authenticate, personalize or bill for such services. Hence, the

SIM-based authentication schemes are good competitors against the Web-based schemes, among other reasons due to the latter does not provide mutual authentication functionality between mobile station and backend server (a client certificate should be required) whereas the SIM-based schemes easily supports such a functionality.

Consequently, the EAP-SIM/AKA standardized protocols along with RADIUS or DIAMETER (supporting AAA procedures) are de facto authentication schemes for the 3G/WLAN interworking architectures [7][12]. By means of a number of proxies, it is possible to transport the authentication messages through a visited wireless local network towards our home 3GPP network, in a roaming situation.

Due to the complexity associated to the network in 3G/WLAN scenarios, most of works have been focused on the security and technical problems in the network side. Thus, some authors highlight the resulting latency during the authentication process and propose techniques based on AAA brokers as third trusted party [8], which manages the security associations and key distribution. Other original works are focused on a proactive key distribution scheme based on a context transfer between foreign and home network [9], and in other cases, as we will see in the section 2 of this paper, they study global security problems associated to the standardized protocols.

Nevertheless, regarding the chip card running in these interworking schemes in a U(SIM) role, few works have been developed and that is the scope of the present paper. More concretely, this paper aims to review the way in which the EAP-SIM/AKA security protocols are used in 3G/WLAN interworking from the point of view of U(SIM), with the goal to provide a more robust and secure solution.

Our new approach starts from a different authentication model [10] that considers an isolated U(SIM) with autonomy during the authentication process. In other words, the U(SIM) participates as stand-alone supplicant or claimant, and not relies on the access terminal (i.e. WLAN mobile station) for this functionality. Additionally, this work assumes an a priori untrustworthy environment, where the WLAN MS is considered as a potential attacker. Hence, the WLAN MS should be authenticated by the network as a different host from U(SIM). Thus, we will define in this paper an AAA architecture, which represents a more robust and flexible solution in terms of security. Beyond these benefits, this approach also provides efficient mobile stations' customization or personalization in critical or public environments.

In the reminder of this paper, the related work is reviewed in section 2 and, afterwards, we describe an AAA architecture based on our network smart card concept [10], NSC, which implements U(SIM) authentication functionalities (NSC-based U(SIM)). In section 4, security and trust issues related to such an architecture are discussed. Finally, we describe the testbed and implementation carried out with the goal to run end-to-end authentication protocols over the proposed architecture and to test her feasibility.

2   Related Work

The advances in the 3G/WLAN interworking systems have been reflected in the standardization process [7][11], where the reference model for different scenarios is detailed. In [12] and other works [13] [14], the security related to theses systems is profusely described. An example of interworking architecture (compounded by UMTS and IEEE802.11 technologies) was evaluated in simulation environments in [15].

After the subscriber authentication phase takes place by means of her U(SIM), the cellular network operator will provide access to certain IP-based services. It is important to highlight the heterogeneity feature in wireless devices and networks under the all-IP concept, which is applied along the end-to-end communication for the provision of multiple services (Web, IMS, VoIP, video streaming, etc.). From the beginning, many works were devoted to this topic. In [16], the call admission control over various DiffServ settings was studied for this kind of architectures and in [17] the session establishment with SIP was tested for the provision of IMS services. In [18] the VoIP throughput into an IPSec tunnel was analysed by forcing the number of connections to an unique access point in mobility situations.

Continuing with the network side, standards and many works have been focused on 3G/WLAN interworking security. The subscriber authentication process (more general, AAA) through the 3G/WLAN architecture in a roaming situation and, obviously, previous to the IP session, is illustrated in Figure 2. A wireless local network based on IEEE 802.11i technology is represented.

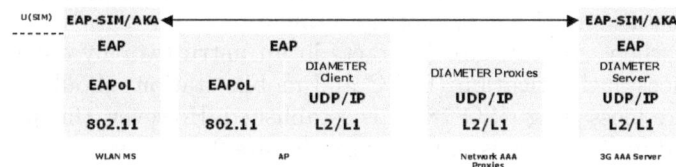

Fig. 2. Example of an AAA protocol architecture in 3G/WLAN interworking

The AAA architecture shown in Figure 2 is based on the EAP-SIM/AKA protocols. In summary, the U(SIM) stores the corresponding subscriber authentication credentials and computes the envisaged cryptographic algorithms in such protocols, on the behalf of mobile station. In order to provide universal support for transmission-level security, and enable both intra- and inter-domain AAA deployments, IPsec support is mandatory in DIAMETER [19][4]. IPsec ESP in transport mode and authentication algorithms provides per-packet authentication, integrity protection, confidentiality and supports replay protection mechanisms.

Nevertheless, some weaknesses in EAP-SIM/AKA schemes have been found [20][21][22]. Since authentication procedure requires multiple request-response exchanges, attacks in visited networks that can compromise authentication vectors in roaming situation have been detected. Moreover, identity privacy is not always guaranteed when the identification of a user is performed by means of the permanent subscriber identity (IMSI) or pseudonym in clear text [23]. Additionally, the system could be actively attacked by a malicious impersonation of the network with the goal to obtain the subscriber's IMSI. Finally, EAP-AKA does not support cipher suite negotiation or protocol version negotiation , therefore negotiation attacks are feasible, as well as, man-in-the-middle attacks.

With the goal of overcoming these flaws, in [21] and more recently in [24] can be found proposals of tunnelled end-to-end authentication schemes based on EAP-TLS [25] or EAP-TTLS [26] over the 3G/WLAN interworking architecture. Other previous working lines, have aimed to make more robust the subscriber authentication and authorization on the basis of temporary attributes certificates [27]. The goal of this proposal was to reduce the inconveniences of the certificates management and their revocation, minimizing the impact on the interworking architecture.

However, the problems derived from the certificate management in the client side or/and from the complexity of tunnels establishing, supported by the U(SIM), suggest to look for more lightweight schemes.

Another problem in the current implementation of EAP protocols in U(SIM) is due to the by default consideration of a implicit trustworthy WLAN MS (e.g. laptop, smartphone, PDA, etc.). That means that both devices blindly trust each other. In fact, they behave as an unique supplicant. In our opinion, this is not a by default recommendable assumption. Thus, the authentication schemes should be designed to protect against any potential scenario, even where the WLAN MS is an a priori untrustworthy terminal.

Moreover, when a smart card interacts in an untrustworthy environment, a previous devices authentication (UICC and mobile station) should be required before a secure messaging (ISO 7816) is established. However, this protection is not considered in [12], as it is illustrated in Figure 2.

Therefore, a more robust approach should be performed in order to obtain versatile solutions. Just note that, an U(SIM) may be an external contact/contacless smart card that customizes (personalizes) a public wireless terminal for a 3G/WLAN access. Specifically in such a case, the U(SIM) behaviour as an stand-alone supplicant is highly recommendable. So it should be isolated and protected. Otherwise, the WLAN MS could be considered as the perfect candidate to be the man in the middle. An example of MitM attack concerning EAP-SIM is described in [28]. This attack breaks the A5/2 algorithm, whenever a few valid GSM triplets have been retrieved.

In the following section, we propose a novel approach on the AAA architecture in Figure 2. This proposal is respectful with the required protocols (EAP-SIM/AKA, RADIUS/DIAMETER, etc.) and it basically aims to improve the robustness and security in this kind of interworking scenarios.

3 New NSC-Based AAA Protocol Architecture in 3G/WLAN

This paper proposes a new AAA protocol architecture for 3G/WLAN infrastructures based on our Network Smart Card concept (NSC-based). Under this scope, we consider an U(SIM) remote authentication scheme, where this device adopts the functionality of stand-alone supplicant instead of split supplicant: the U(SIM) and WLAN MS does not cooperate in the authentication process as an unique device. That is why, in our work, the authentication protocol stack is designed as an integral part of the U(SIM) (atomic design). With this goal, we propose a specific protocol stack for the chip card that participates as actual endpoint in the authentication process with a 3G AAA server.

This new architecture (Figure 3) implies minimal changes in the original one (Figure 2) but it introduces significant advantages. For instance, in the 3G network side no changes are needed. Thus, proxies and end-equipments keep settings and implementation features.

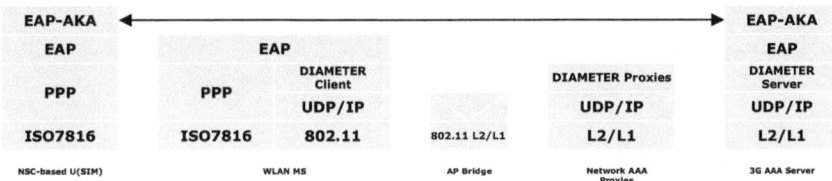

Fig. 3. Our NSC-based AAA protocol architecture in 3G/WLAN interworking

The proposed AAA protocol architecture requires a simpler protocol implementation for the WLAN acces points (APs) with U(SIM) remote authentication purposes. Note that WLAN Mobile Station participates as a Network Access Server (NAS) implementing the role of pass-through authenticator as a DIAMETER client according to [4].

In a first phase, the DIAMETER server authenticates the WLAN MS by her own mechanisms. In a second phase, the function of the pass-through authenticator is shifted to WLAN MS. This reinforces the stand-alone supplicant functionality in the U(SIM), since WLAN MS cannot act as supplicant and authenticator at the same time for the same U(SIM). One should note the advantages that the U(SIM) isolation brings with regard to assure the security of the entire scheme in untrustworthy scenarios.

Our architecture takes advantage of the functions of the LCP protocol that is provided by PPP [29]. LCP/PPP protocol may be easily hosted in the U(SIM) stack. The functions for controlling network included in the NCP sub-protocol are beyond the scope of this work. On the other hand, PPP offers versatility in the authentication, thanks to its extensibility. In fact, EAP (Extensible Authentication Protocol) was initially designed for PPP. According to our approach, the EAP Layer must be atomically implemented in the smart card and must allow the packets exchange between the EAP-SIM/AKA methods and LCP frames, as well as, the duplication and retransmissions control.

Based on this architecture, an authentication messages exchange has been designed in our work. Figure 4 illustrates this authentication flow.

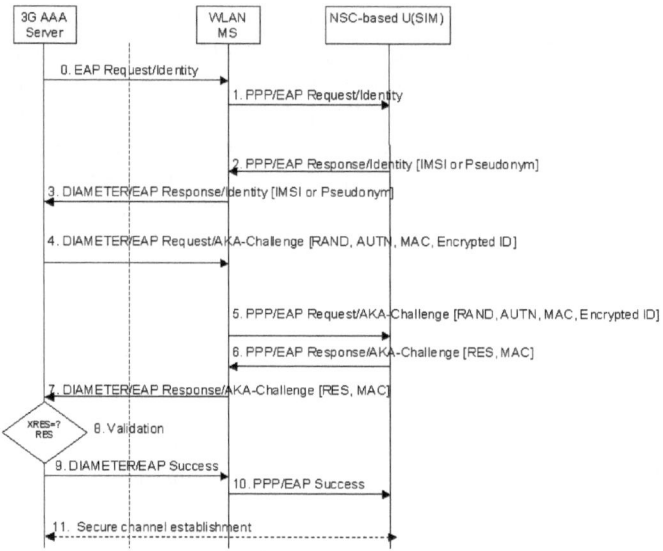

Fig. 4. Authentication Flow in our AAA architecture

The NSC-based U(SIM) authentication process is as follows:

1. The WLAN MS (representing the network and providing WLAN access) sends an PPP-EAP request identity (either an IMSI or a pseudonym) message to the NSC-based U(SIM) in order to initiate the procedure.

2. The NSC-based U(SIM) returns the EAP Response/Identity packet to the WLAN MS.

3. The WLAN MS sends the EAP Response/Identity packet to the 3G AAA Server in network. The authentication messages exchange between WLAN MS and 3G AAA Server are encapsulated into DIAMETER packets.

4. The 3G AAA Server initiates the EAP AKA authentication process with the appropriate EAP Request/AKA-Challenge message.

5. The WLAN MS processes the DIAMETER headers and sends the received EAP packet to the NSC-based U(SIM), encapsulated into a PPP frame.

6. The NSC-U(SIM) returns the EAP Response/AKA-Challenge packet to the 3G AAA Server, which will check the validity of the RES.

7. The WLAN MS builds the corresponding DIAMETER packet and sends it to the 3G AAA Server.

8. The 3G AAA Server checks the validity of the RES and computes the MAC of the entire received message, and she compares it with the received MAC.

9. In case of a correct validation, the NSC-based U(SIM) is authenticated and the 3G AAA Server sends an EAP Success packet to the NSC-based U(SIM).

10. The WLAN MS retransmit the EAP Success packet on the PPP link.

11. After a successful EAP authentication, the NSC-based U(SIM) is authorized by the network equipment (e.g., WLAN MS or even the actual 3G AAA server). Both devices could derive/know a master session keys to establish a secure channel (secure messaging) between them.

As is stated in [7], an EAP-AKA fast re-authentication procedure was developed with the goal to make more lightweight the authentication process. Note that the EAP-AKA authentication process may be frequently performed in order to obtain fresh authentication vectors from the home network. By means of fast re-authentication procedure, the certain keys that have been derived in a previous full authentication are reused, so just one new master session key is generated with link layer protection purposes. The inclusion of the EAP-AKA fast re-authentication in our scheme is trivial.

4   Security and Trust Issues

Regarding the security aspects of our architecture, it should be noted that we are not proposing a new U(SIM) authentication protocol in the context of 3G/WLAN interworking. Our architecture is designed by well-known protocols that are implemented inside the U(SIM) with a novel approach.

Nevertheless, this new architecture determines a new way to transport authentication messages between the U(SIM) and a 3G AAA server, and where the U(SIM) takes the control in the user side. Therefore, the security weakness and threats are derived by the own nature of such standardized protocols and the correctness of their implementation.

Additionally, new secure algorithms, key material or cryptographic techniques are not required. The implementation of the EAP-SIM or EAP-AKA methods is transparently reused, both in the U(SIM) side and in the 3G AAA Server side. However, one of the more important impacts of our proposal is related to the trust models. If we study the trust model, Figure 5, derived from the current AAA protocol architecture in a 3G/WLAN interworking scenario (Figure 2), we

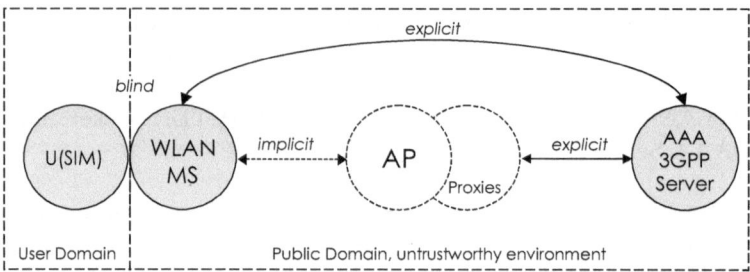

Fig. 5. Trust model in the original architecture

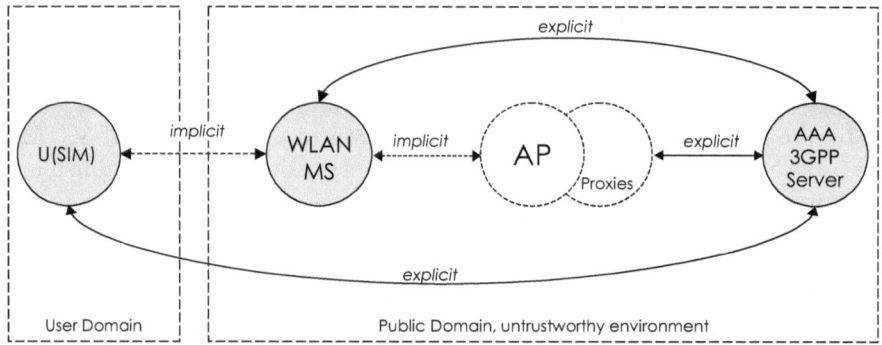

Fig. 6. Trust model in our architecture

observe that there exists an explicit trust between AP and 3GPP AAA server (supported by DIAMETER protocol) and an explicit trust between WLAN MS and 3GPP AAA server after a successful authentication process (supported by an EAP method). In any case, the trust relationship in the interface between U(SIM) and WLAN MS is not questioned and it could be considered as "blind". As we mentioned before, this assumption should not be applied to all scenarios and a more flexible solution is required. With this goal, we introduce a more realistic architecture, which a new trust model is derived from, Figure 6.

In our trust model, the trust relationship between the WLAN MS and the 3G AAA server is supported by DIAMETER protocol (e.g pre-shared keys) and such trust relationship could be considered as explicit. Here, the WLAN MS is part of the network and it behaves as an access point for the U(SIM). The trust relationship between U(SIM) and WLAN MS is a priori null (untrustworthy). After an end-to-end successful authentication process (supported by an EAP-SIM/AKA method) between the U(SIM) and 3G AAA Server, the trust relationship between them should be now considered explicit, as result of a mutual authentication process. Therefore, in this point the trust relationship between U(SIM) and WLAN MS is just implicit, since no direct mutual authentication process between them has occurred. In other words, just when U(SIM) trusts 3G AAA server then she trusts WLAN MS. This is a reasonable result in a priori untrustworthy scenarios.

Moreover, in untrustworthy scenarios a device authentication process should occur, i.e. an authentication process between devices based on shared keys (or card verifiable certificates), directly driven by the involved devices. In this context, it does not make sense to perform two mutual remote authentication, i.e. subscriber authentication and device authentication, so a local device authentication (U(SIM)- WLAN MS) may take place in order to avoid an additional management of the key material in a number of public WLAN MSs. By means of our AAA protocol architecture the corresponding master session key (also derived by U(SIM)) is sent to WLAN MS from the network side. Therefore, the (U)SIM-WLAN MS interface is per-session authenticated and protected against potential attacks (e.g. MitM attack, WLAN MS impersonation).

Although some flaws in EAP-AKA have been proved by several authors, the tunnel-based solutions (e.g. based on EAP-TLS) are interesting proposals, which could deal with these weaknesses. In principle, our architecture could further implement this kind of protocols, though performance tests should be carried out.

5   Implementation and Testbed

The testbed for the AAA network architecture is represented in Figure 7. It has been implemented by means of the OpenDiameter [30] libraries. OpenDiameter libraries provide a C++ API both to EAP and Diameter EAP.

3G AAA Server

The back-end authentication server is basically implemented in a computer by the *libdiametereap* library. Such a library implements the specification defined in [4]. The Diameter EAP API is extensible in a way that server applications can define its own authorization decisions for each authorization attribute carried in Diameter EAP Answer (DEA) messages.

Additionally, the *libeap* library implements a set of state machines of EAP, which is specified in [31] . In this case, this library provides an EAP backend authenticator implementation.

The EAP API is extended in order to support EAP-AKA as a new authentication method including the corresponding method's state machine and message parsing. On the other hand, the *OpenSSL* library includes a general purpose cryptography library, which is partially included in this testbed with the goal of providing a set of AKA cryptographic functionalities. Since this work is focalised on authentication purposes, for simplicity's sake, the implementation of functions f3 and f4 [32] has not been carried out. This functions are envisaged with key agreement purposes (CK and IK). These keys would be used to derive further keying material with different goals: e.g. EAP-AKA additional packets protection, link layer security, in HMAC algorithm or fast re-authentication identity encryption.

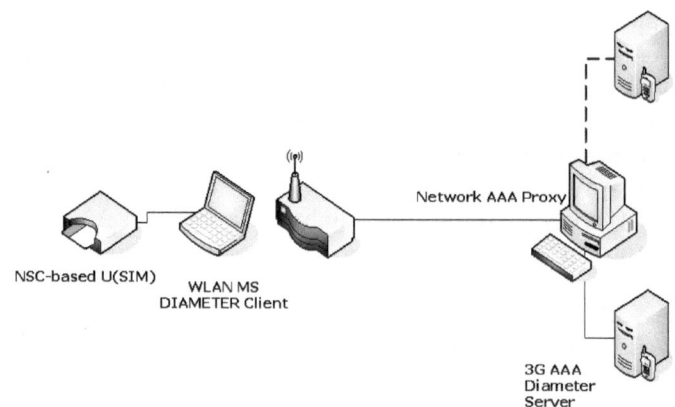

Fig. 7. Testbed for our architecture

Network AAA proxy

Multiple network AAA proxies could intermediate between the wireless LAN network and the 3G network. Our testbed considers just one proxy, which simulates one of these entities. The standard Diameter base protocol procedure in her relay version (Diameter proxy) is provided by the *libdiameter*. It allows us to complete the implementation of the adequate protocol stack in a layer 2 wireless Access Point. In our testbed, Diameter messages are hop-by-hop protected by IPsec with pre-shared keys (IKE Aggressive Mode) between WLAN MS (NAS), AAA proxy and between this one and AAA server.

WLAN MS

The WLAN mobile station is a common laptop with a IEEE 802.11g wireless interface. The functionality of NAS (Diameter client) is provided by the implementation of the *libdiametereap* library.

Network Smart Card with U(SIM) functionalities

The base implementation in the smart card for this testbed is previously described in [10]. Thus, the bulk LCP/EAP protocol stack -according to the standardized state machines- has been enhanced with a set of functionalities corresponding EAP-AKA method. As is stated before, CK and IK derivation, as well as, synchronization and re-authentication functionalities have been avoided with testbed experiments purposes. Partial view of the EAP- AKA state machine is illustrated in Figure 8.

Although we are continually improving the implementation of this architecture and protocol, we have measured a initial performance time of 6-7 sec. for completing the authentication process (authorization policy is excluded) in laboratory environment. The Sm@rtCafé Expert 3.x and Sm@rtCafé Expert 64 smart cards and G&D's development tools [33] have been used for the experiments in our testbed.

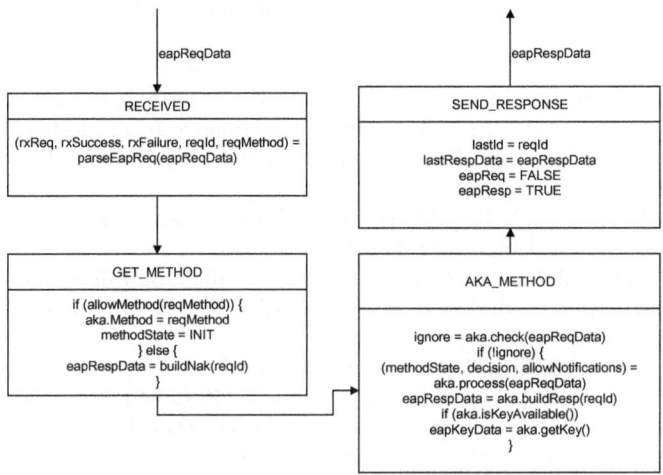

Fig. 8. Partial view of the EAP state machine in the smart card

6   Conclusion

Our testbed shows the feasibility and robustness of the proposed NSC-based AAA protocol architecture for 3G/WLAN interworking scenarios. The standardized EAP-AKA protocol is transparently implemented in a common U(SIM), which participates as stand-alone supplicant (NSC-based U(SIM)), and she does not rely on the WLAN mobile station for this functionality. This feature defines a novel trust model that assumes an a priori untrustworthy environment, where the WLAN MS is considered as a potential attacker. Thus, our approach represents a more flexible solution in terms of security. Beyond these benefits, it also provides efficient mobile stations' customization or personalization in critical or public environments. Next future works will study other related protocols over the same architecture and they will in depth treat performance tests.

Acknowledgements. This work is supported by ASPECTS-m Project, CICYT-2004-SEG-04000.

References

1. Leu, J.-S., Lai, R.-H., Lin, H.-I., Shih, W.K.: Running cellular/PWLAN services: practical considerations for cellular/PWLAN architecture supporting interoperator roaming. IEEE Communications Magazine 44(2), 73–84 (2006)
2. Rigney, C., Willens, S., Rubens, A., Simpson, W.: Remote Authentication Dia. In: User Service (RADIUS), IETF RFC 2865 (June 2000)
3. Aboba, B., Calhoum, P.: RADIUS (Remote Authentication Dial In User Service) Support For Extensible Authentication Protocol (EAP), IETF RFC 3579 (September 2003)

4. Eronen, P., Hiller, T., Zorn, G.: Diameter Extensible Authentication Protocol (EAP) Application, IETF RFC 4072 (August 2005)
5. Haverinen, H., Salowey, J.: Extensible Authentication Protocol Method for GSM Subscriber Identity Modules (EAP-SIM), IETF RFC 4186 (January 2006)
6. Arkko, J., Haverinen, H.: Extensible Authentication Protocol Method for 3rd Generation Authentication and Key Agreement (EAP-AKA), IETF RFC 4187 (January 2006)
7. 3GPP TS 23.234 v7.3.0: 3GPP System to Wireless Local Area Network (WLAN) Interworking System Description (September 2006)
8. Salgarelli, L., Buddhikot, M., Garay, J., Patel, S., Miller, S.: Efficient authentication and key distribution in wireless IP networks. IEEE Wireless Communications 10(6), 52–61 (2003)
9. Shin, M., Ma, J., Mishra, A., Arbaugh, W.A.: Wireless network security and interworking. Proceedings of the IEEE 94(2), 455–466 (2006)
10. Torres, J., Izquierdo, A., Sierra, J.M.: Advances in network smart cards authentication. Computer Networks 51(9), 2249–2261 (2007)
11. Ahmavaara, K., Haverinen, H., Pichna, R.: Interworking architecture between 3GPP and WLAN systems. IEEE Communications Magazine 41(11), 74–81 (2003)
12. ETSI TS 133 234 V7.5.0, 3GPP System to Wireless Local Area Network (WLAN) Interworking Security System (June 2007)
13. Koien, G.M., Haslestad, T.: Security aspects of 3G-WLAN interworking. IEEE Communications Magazine 41(11), 82–88 (2003)
14. Salkintzis, A.K.: Interworking techniques and architectures for WLAN/3G integration toward 4G mobile data networks. IEEE Wireless Communications 11(3), 50–61 (2004)
15. Siddiqui, F., Zeadally, S., Yaprak, E.: Design Architectures for 3G and IEEE 802.11 WLAN Integration. In: Lorenz, P., Dini, P. (eds.) ICN 2005. LNCS, vol. 3421, pp. 1047–1054. Springer, Heidelberg (2005)
16. Song, W., Jiang, H., Zhuang, W., Shen, X.: Resource management for QoS support in cellular/WLAN interworking. IEEE Network 19(5), 12–18 (2005)
17. Marquez, F.G., Rodriguez, M.G., Valladares, T.R., de Miguel, T., Galindo, L.A.: Interworking of IP multimedia core networks between 3GPP and WLAN. IEEE Wireless Communications, [see also IEEE Personal Communications] 12(3), 58–65 (2005)
18. Rajavelsamy, R., Jeedigunta, V., Holur, B., Choudhary, M., Song, O.: Performance evaluation of VoIP over 3G-WLAN interworking system. IEEE Wireless Communications and Networking Conference 4(13-17), 2312–2317 (2005)
19. Calhoun, J., Loughney, E., Guttman, G., Zorn, J.: Arkko, Diameter Base Protocol, IETF RFC 3588 (September 2003)
20. Barkan, E., Biham, E., Keller, N.: Instant Ciphertext-Only Cryptoanalysis of GSM Encrypted Communication. In: Crypto 2003 (August 2003)
21. Kambourakis, G., Rouskas, A., Kormentzas, G., Gritzalis, S.: Advanced SSL/TLS-based authentication for secure WLAN-3G interworking. IEE Proceedings Communications 151(5), 501–506 (2004)
22. Meyer, U., Wetzel, S.: A man-in-the-middle attack on UMTS. In: Proceedings of the 2004 ACM Workshop on Wireless Security, October 2004, pp. 90–97 (2004)
23. Cheng, R.G., Tsao, S.L.: 3G-based access control for 3GPP-WLAN interworking. In: IEEE 59th Vehicular Technology Conference, VTC 2004-Spring, May 2004, vol. 5, pp. 2967–2971 (2004)

24. Zhao, Y., Lin, C., Yin, H.: Security Authentication of 3G-WLAN Interworking. In: 20th International Conference on Advanced Information Networking and Applications. In: AINA 2006, April 2006, vol. 2, pp. 429–436 (2006)
25. Aboba, B., Simon, D.: PPP EAP TLS Authentication Protocol, IETF RFC 2716 (October 1999)
26. Funk, P., Blake-Wilson, S.: EAP Tunneled TLS Authentication Protocol Version 1, (EAP-TTLSv1), Internet-Draft, draft-funk-eap-ttls-v1-01.txt (March 2006)
27. Kambourakis, G., Rouskas, A., Gritzalis, S., Geniatakis, D.: Support of Subscribers Certificates in a Hybrid WLAN-3G Environment. Computer Networks 50(11), 1843–1859 (2006)
28. Barkan, E., Biham, E., Keller, N.: Instant Ciphertext-Only Cryptoanalysis of GSM Encrypted Communication. In: Boneh, D. (ed.) CRYPTO 2003. LNCS, vol. 2729, pp. 600–615. Springer, Heidelberg (2003)
29. Simpson, W.: The Point-to-Point Protocol (PPP), IETF RFC 1661, Standard Track (July 1994)
30. Open Diameter Project. Open-source software for the Diameter base protocol and others, http://www.opendiameter.org/
31. Vollbrecht, J., Eronen, P., Petroni, N., Ohba, Y.: State Machines for Extensible Authentication Protocol (EAP) Peer and Authenticator, IETF RFC 4137 (August 2005)
32. 3GPP TR 35.909 V7.0.0, Technical Specication Group Services and System Aspects; 3G Security; Specication of the MILENAGE Algorithm Set: An example algorithm set for the 3GPP authentication and key generation functions f1, f1*, f2, f3, f4, f5 and f5*; Document 5: Summary and results of design and evaluation (Release 7) (June 2007)
33. Sm@rtCafé Professional Toolkit 2.0, G&D, http://www.gi-de.com

Using TPMs to Secure Vehicular Ad-Hoc Networks (VANETs)

Gilles Guette[1] and Ciarán Bryce[2]

IRISA
Campus de Beaulieu, 35042 Rennes CEDEX, France
gilles.guette@univ-rennes1.fr, Ciaran.Bryce@inria.fr

Abstract. Vehicular Ad hoc Networks are the focus of increased attention by vehicle manufacturers. However, their deployment requires that security issues be resolved, particularly since they rely on wireless communication, and rogue vehicles can roam with contaminated software. This paper examines security threats to VANETs and argues that a security architecture built around TPMs can provide a satisfactory solution.

1 Introduction

Over the last few years, Vehicular Ad hoc Networks (VANETs) have gained much attention within the automobile and research worlds. One reason is the interest in a growing number of applications designed for passenger safety – such as emergency braking, traffic jam detection and cooperative driving – as well as in applications aiming at the comfort of passengers, such as games, chat-rooms and vehicle data-sharing (e.g., CarTorrent [1]).

VANETs are highly dynamic *ad hoc* networks of devices with very restricted access to a network infrastructure. Moreover, if base stations are sparsely deployed along the road, access is also of short duration due to vehicle speed. Since on-board applications need to exchange data, the communication security problem must be addressed. The absence of a permanently present infrastructure means that a decentralized security architecture is required. Given the safety critical nature of some VANET applications, the security architecture must imperatively prevent a malicious person from successfully launching an attack intending to provoke collisions between vehicles.

This paper examines some of the security requirements for VANETs for two selected applications – platoons and event reporting. The security analysis insists on the need for pseudonymity, trustworthy information exchange and a fail-safe mode where doubts over the trustworthiness of information can be conveyed to the vehicle (driver). We contend that these requirements can be met in a scalable way by a security architecture built around the emerging Trusted Platform Module (TPM) [2] specification. This is partly because use of the TPM makes it easier to verify that correct functioning of software on a vehicle, and the distribution of keys for TPM operation can be accommodated by current vehicle registration and maintenance practices.

J.A. Onieva et al. (Eds.): WISTP 2008, LNCS 5019, pp. 106–116, 2008.

The remainder of this paper is organized as follows. Section 2 presents related work. Section 3 presents two example VANET applications along with their security requirements. The TPM is presented in Section 4, and the TPM based security solution is outlined in Section 5. Section 6 concludes the paper.

2   Related Work

In a wireless Vehicular Ad hoc Network, as data is broadcast over a shared communication media, it is simple for a malicious node to intercept or modify data, or to inject erroneous data. A data injection can provoke collisions in a vehicular platoon [3]. The open nature of a VANET thus renders communication security a great challenge [4,5,6,7].

One approach to VANET security has been to adopt a VANET PKI (VPKI) [8,9] that allows vehicles to securely communicate among themselves. Base stations placed along the road provide support for the infrastructure, notably for key distribution and revocation.

VPKI solutions address privacy using an *anonymous key set* and a *key changing algorithm* to avoid the possibility of car tracking. Without key changing, a vehicle would use the same public key to sign all of its messages. It would thus be easier for an eavesdropper on the network to correlate the vehicle's positions with the public key holder.

VPKIs are promising for VANET applications. However, the PKI deployment is a large-scale and potentially costly procedure since it requires large-scale testing after deployment to ensure operation under real-world VANET conditions. Further, the solution really only aims at ensuring authentication (of pseudomyns). As we argue in the next section, a security infrastructure must be aimed at establishing the authenticity of message contents for safety and security.

Some other papers address the problem of privacy [10,11,12] in VANET with the help of infrastructures (base station and certification authorities) and pseudonym use. [12] deals with the challenges encountered when applying anonymity to a VANET communication system and proposes a framework for pseudonymity support. A study of the impact of pseudonym changes on geographic routing in VANETs is made in [13]. All of these papers underline that supporting pseudonymity requires changing other *identifiers* of the protocol stack, such as IP or MAC addresses.

3   Use Case: Cooperative Driving

In this section, we present two cooperative driving applications for Vehicular Ad hoc Networks. This is the subject of Section 3.1. Section 3.2 then presents the security threat model for these applications, and Section 3.3 finishes with a list of desired security properties.

In the context of this paper, vehicles are nodes of an *ad hoc* network, and we use the terms *node, car* and *vehicle* interchangeably. Each car has wireless networking capabilities (e.g., ad hoc WLAN) and possesses a GPS device for positioning itself. A vehicle may have further sensor devices, e.g., for sensing the weather conditions. The set of sensors maintained by a vehicle is termed its *configuration*.

Note, we do not expect each car circulating on the road to be part of a VANET. Rather, VANET functionality will be something progressively introduced into new cars. Even then, there will always be old model cars as well as foreign vehicles on the roads. Thus, the cooperative driving use cases do not rely on all cars being VANET nodes.

3.1   Description

Each node has embedded sensors to detect environment information. While the sensor configuration may differ from one node to another, all nodes of the VANET use wireless communication to broadcast and share information obtained via their sensors about the state of the traffic – traffic jams, road fluidity, obstacles, weather conditions, *etc*. As suggested in Figure 1, information sharing may help to avoid accidents by enabling drivers to adapt their behavior based on pertinent safety information from vehicles driving in the opposite direction.

Another cooperative driving application, based on inter-vehicle information sharing information, is vehicular platoons [3], *c.f.*, Figure 2. A platoon reduces the distance between vehicles, and this has the economic and ecological advantage of reducing fuel consumption. The application embedded in the vehicle manages the distance between a vehicle and its predecessor and successor vehicles, and manages variations in speed.

Both of these applications rely on the vehicles' configuration returning accurate readings. For instance, the GPS module of a given vehicle can make an error in positioning, or its internal clock may be erroneous, thus indicating an event that is more recent than its message suggests. Further, every car is different – the time it takes to accelerate or decelerate is different for each car and this has an important impact in the platoon application. It is therefore important for the vehicle's configuration to be up-to-date with respect to device and sensor characteristics: the vehicle must continuously measure its configuration so that other nodes can interpret messages received from it with respect to sensor accuracy.

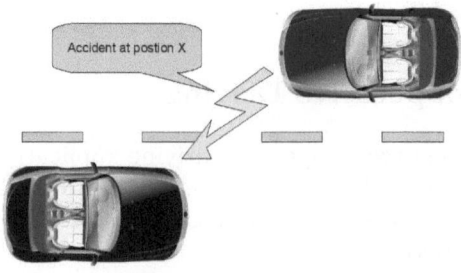

Fig. 1. Data transmission

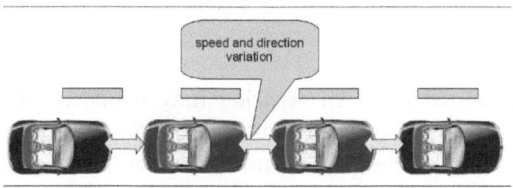

Fig. 2. Vehicle platoons

3.2   Threat Model

As mentioned, the reception of traffic information can modify the behavior of a driver. When a message announces an obstacle on the road or fog a few kilometers ahead, the control screen of the vehicle or a digital voice alerts the driver. There is a high probability that the driver slows down.

The reduction of distances between vehicles increases the risk of collisions if an attacker can send wrong information to a vehicle at the wrong moment. We need to avoid such a possibility because this kind of attack can have catastrophic consequences. Generally, the security architecture has to deal with the following attacks.

The Sybil Attack. The Sybil attack was first described and formalized by Douceur in [14]. It consists of a node sending multiple messages, with each message containing a different, fabricated, source identity. Thus, the attacker appears in the network as a large number of different nodes. Applications of the Sybil attack to Vehicular Ad-Hoc Networks are discussed in [3,15] and show the importance of Sybil node detection in VANETs. A possible goal of a Sybil attack by an attacker is to give the illusion to other cars that there is a traffic jam and thus encourage other vehicles to leave the road to the attacker's benefit. Nevertheless, this attack may be more dangerous, targeting directly human life by trying to provoke collisions [3].

One important result shown in [14] is that without a logically centralized authority, Sybil attacks are always possible (*i.e.* may remain undetected) except under the extreme and unrealistic assumption of resource parity and coordination among entities.

Node Impersonation. Drivers are legally responsible for their actions behind the wheel. In the event of an accident or a driving offense, there is a need for the police to associate the implicated vehicle with its driver. This is currently possible thanks to databases of driving license plates. In a VANET, this can be easily accomplished by giving a unique identifier to every vehicle. In case of an accident provoked by wrong information sent by a vehicle, the message can be verified and its identifier controlled. The police may then bind the identifier with the driver's identity.

This identifier must be protected so that an attacker cannot masquerade with a fabricated or some other car's identity. At the same time, for privacy reasons, it must not be possible for someone to deduce the driver's identity from the vehicle's VANET identifier.

Sending False Information. An attacker may want to send wrong or forged data to other vehicles to provoke collisions, to free the road or for some other goal. This threat against the vehicle may be mitigated by the fact that there exists a way to know the identifier of the sender of a message. Nonetheless, the security mechanism must integrate ways to estimate the truthfulness of information.

Car Tracking. Driver privacy is a concept that must be integrated into the security solution. Drivers may wish to preserve their anonymity even though the use of unique identifiers allows vehicles to be tracked. Nevertheless, in [6] the authors underline an important fact: today vehicle are only partially anonymous. Drivers implicitly surrender a portion of their privacy since each vehicle has a publicly displayed license plate that uniquely identifies it. It might not be difficult to link a license plate to the driver's identity.

The use of wireless communication does not add a new problem threatening the driver's privacy. Nevertheless, as data is broadcasted over a potentially long range, it becomes easier to collect data. Moreover, if base stations are deployed along the road, data might be collected by a third party with a commercial aim. Solutions based on the use of pseudonyms are presented in [12,10,11]

3.3   Basic Security Properties

Regarding the different threats exposed in the previous section, we can define basic properties that a security solution must provide.

Property 1. A node must have a unique identifier. This identifier may be associated with a set of pseudonyms, but in this case an authority must have the possibility of linking a given pseudonym to its associated unique identifier.

Property 2. To avoid modification of a given message or a wrongful claim of identity in a message, each message must be authenticated with regards to a vehicle identifier, and the integrity of this message must be ensured.

To engage the liability of a driver having caused an accident, non-repudiation must also be provided by the security solution. Nevertheless, this security property is implicitly provided by the combination of properties 1 and property 2 – if a message containing the unique identifier of its sender cannot be modified, then non-repudiation is effectively provided.

Property 3. The trustfulness of message contents must be verifiable.

This is a stronger property than before. In effect, it entails being able to challenge a vehicle for it to prove that its configuration readings are correct. In effect, the security infrastructure is more linked to demonstrating correct functionality rather than identity.

One way to avoiding false information exchange is to authenticate the application, rather than the vehicle, sending the information. If we can prove that information have been sent by a cooperative driving application and that this application has no been hacked, we can ensure that this information is not voluntarily wrong.

Avoiding information that is involuntary wrong, e.g., due to erroneous sensor readings, is achieved by challenging vehicles for readings while in the same geographical vicinity. In this case, the readings of the challenger and the challenged vehicle should not significantly diverge. If they do, then either vehicle has an error and should avoid a platoon with other vehicles.

4   The Trusted Platform Module

Implementing the security properties that we presented in Section 3 requires that a vehicle be able to establish trust in another vehicle, even though that vehicle is under the complete control of an unknown, and therefore untrusted, driver. The solution thus requires the use of secure hardware. An example of a general purpose hardware chip designed for secure computing is the *Trusted Platform Module* (TPM) [2] which can be integrated into any device. TPMs are now shipped with PCs; 200 million TPM-enabled PCs have been shipped by the end of 2007.

A TPM is a piece of hardware, requiring a software infrastructure, that is able to protect and store data in shielded locations. A TPM has also cryptographic capabilities such as a SHA1 engine, an RSA engine and a random number generator. Figure 3, taken from [2], illustrates the main components of a TPM.

A trusted platform must provide three basic features: protected capabilities, integrity measurement and integrity reporting. Integrity measurement and reporting mean that at the first boot of a platform, the TPM is able to store a fingerprint of application and environment variables in a specific shielded location called a Platform Configuration Register (PCR). In principle, any change

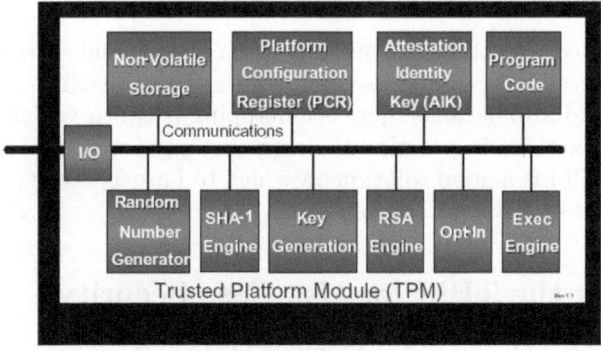

Fig. 3. Architecture of a TPM

that an attacker tries to make to the application will lead to a difference between the original fingerprint of the application and the fingerprint of the hacked application, thus allowing other devices to detect that a hacked application is being run.

The data used to take fingerprints are stored by the platform in a Stored Measurement Log (SML); only the digests of this data are stored in PCRs. During a challenge, the challenger requests to see specific PCR values. Then, an agent on the platform collects the SML entries and receives the PCR values from the TPM. The challenged TPM signs the PCR values with an *attestation identity key* (AiK). The platform agent collects certificates, or credentials in TPM parlance, the signed PCR values, SML entries and returns these to the challenger. Finally, the challenger verifies all the received elements. This procedure is known as an *attestation* protocol.

The credentials involved in attestation serve to demonstrate that the TPM is operating correctly and that the AiK was generated by a valid TPM. There are several keys and credentials, the most important being:

- Each TPM has a unique master key called an Endorsement Key (EK). This is a pair of RSA keys with a minimum size of 2048 bits. Storing this key inside the TPM ensures its security. The public part of the EK is available in the Endorsement Credential. This credential is available outside the TPM itself. The EK is generated by the TPM constructor.
- A Platform Credential is created and signed by the platform provider in which the TPM is integrated and identifies the platform. Generally the platform provider, or some entity that he trusts, will test the TPM and issue a Conformance Credential for the TPM. The conformance credential proves that the TPM has passed the different phases of evaluation.
- A TPM can generate AiKs for attestation protocols. However, credentials must also be issued for these keys that certify that the TPM that generated the key is valid. The TPM specification describes a protocol where a trusted party known as a *privacy CA* can generate AiK credentials for TPMs. The advantage of this is that AiKs credentials need not disclose platform identity in an attestation protocol.

The TPM embedded also offers the possibility of creating and storing encryption keys for data. This feature can be used by vehicles to store driver data securely. For this reason, and also because there is a mapping between vehicle registration and review organizations in practice allowing the appointment of privacy CAs, means that the TPM is a good solution on which to base a security architecture. This is the subject of the next section.

5   Exploiting the TPM for Use Case Security

The security model works at two levels. The basic level permits a *trusted channel* to be established between any two vehicles. This means that the two vehicles are satisfied that each is running an untampered version of the security software, and that no intentional data attack or Sybil attack is being attempted. The second level aims at *information verification*. It builds on trusted channels to offer means to ensure that a vehicle's configuration does not contain erroneous readings.

Implementing trusted channels relies directly on the TPM's attestation mechanism. A vehicle can trust another if the latter can demonstrate that its software has not been tampered with, and the source of the software can be verified. The issue in deploying a TPM on VANET nodes is to assign roles to the actors in the TPM protocols. We assume the following:

- Car manufacturers sign the platform credentials for their vehicles. It is logical to assume that a manufacturer takes responsibility for all embedded devices on their vehicles. Further, manufacturers are relatively few in number and are "well-known" in the sense that certificates signed by these principals should be recognizable to all vehicles and automobile authorities.
- Automobile authorities are responsible for organizing technical reviews. In most countries, car owners are obliged to submit their car to a technical review every 2 to 3 years. A car that fails the technical review cannot be driven on the road. Automobile authorities are thus well-known principals that can act as privacy CAs that can sign AiK credentials.

The TPM provides a means to securely attribute a vehicle identifier. This can be signed by an automobile authority, thus ensuring Security Property 1 of Section 3. The attestation protocol used when vehicles exchange information then ensures Security Property 2.

The second level of security, information verification, is based on three simple procedures. These guarantee Security Property 3.

1. *Auto-measuring.* A vehicle's software maintains data on the vehicle's acceleration and deceleration capabilities, as well as related data such as tire denseness (which embedded devices are now able to measure). These values evolve so the vehicle continuously updates them. These values are obviously important for the platoon scenario where neighboring cars need to agree on minimal distances.
2. *Challenge-response* protocol. This procedure is needed to detect unintentional errors in information transmitted by a vehicle that are due to permanent errors in the sensor of the car. Cars that are close together should possess the same readings for many information types, e.g., temperature, time, location, luminosity. The goal is thus to permit a car to challenge another with respect to any of these readings.
3. *Technical review.* Technical reviews – organized by automobile authorities – for cars with VANET functionality must include reviews of the correct functioning of all sensor devices. Further, we expect that any changes that need to be made to the application software is made at this moment. This is important since the TPM can only be used to help verify that the software on a platform has not been tampered with; in no way does this guarantee that absence of security flaws or bugs in the software itself.

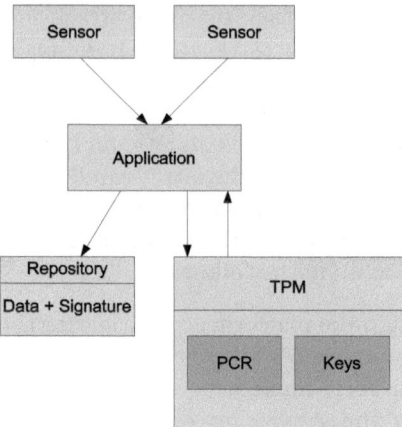

Fig. 4. The embedded architecture

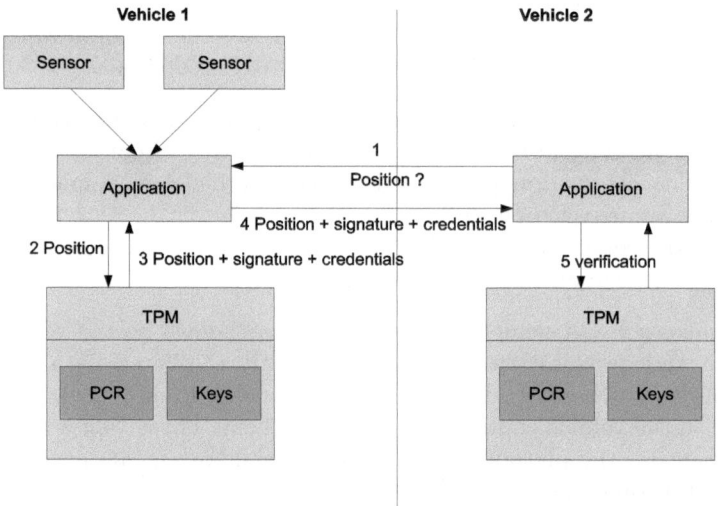

Fig. 5. The challenge-response protocol

The three procedures help to detect and isolate permanent errors in readings. Obviously, intermittent errors are not necessarily treated, and we will look more into this issue in future work. However, we note that these errors are especially a problem for the information exchange scenario of Figure 1 and less so for the platoon scenario (of Figure 2). The latter is more safety critical.

Figure 4 shows the different components of the embedded architecture and the data flow. For instance, for auto-measuring, sensors embedded in vehicle give results of their measures to the application. Then the application asks the TPM to sign the data, the TPM checks the PCR value associated with this application and signs data provided by the application. Then the application can store this data in a dedicated repository.

The details for challenging another vehicle in order to detect unintentional errors is given in Figure 5. The challenger sends a query about data it can verify, the current position in the exemple. Then the challenged vehicle collects the appropriate data, gives this data to its TPM. The TPM checks the PCR values associated with this application and signs data. The application sends to the challenger the signed data and associated credential. The challenger verifies the signature and then can compare the given position to its own current position to detect misconfiguration of the positioning unit of the challenged vehicle.

6   Conclusion and Future Work

In this paper, we have presented the benefit provided by the TPM architecture in Vehicular Ad hoc Networks. We described an application of cooperative driving and its associated threat model. We claim that an embedded TPM inside vehicles can greatly increase the security of wireless communication in this kind of network, and serves as a basis for detecting both intentional and accidental attacks. We are currently working on improvements of our model, notably to the way that updates to application code and embedded certificates are handled.

References

1. Lee, K.C., Lee, S.-h., Cheung, R., Lee, U., Gerla, M.: First Experience with CarTorrent in a Real Vehicular Ad Hoc Network Testbed. Mobile Networking for Vehicular Environments, 109–114 (2007)
2. Trusted Computing Group: TPM main specification. Main Specification Version 1.2 rev. 85, Trusted Computing Group (February 2005)
3. Blum, J., Eskandarian, A.: The Threat of Intelligent Collisions. IT Professional 6(1), 24–29 (2004)
4. Zarki, M.E., Mehrotra, S., Tsudik, G., Venkatasubramanian, N.: Security Issues in a Future Vehicular Network. European Wireless (2002)
5. Hubaux, J., Čapkun, S., Luo, J.: The Security and Privacy of Smart Vehicles. IEEE Security and Privacy 2(3), 49–55 (2004)
6. Parno, B., Perrig, A.: Challenges in Securing Vehicular Networks. In: Fourth Workshop on Hot Topics in Networks (2005)
7. Aijaz, A., Bochow, B., Dötzer, F., Festag, A., Gerlach, M., Kroh, R., Leinmüller, T.: Attacks on Inter-Vehicle Communication Systems - An Analysis. In: 3rd International Workshop on Intelligent Transportation (2006)
8. Raya, M., Hubaux, J.: The Security of Vehicular Ad Hoc Networks. In: Proceedings of the 3rd ACM workshop on Security of ad hoc and sensor networks, pp. 11–21 (2005)
9. Raya, M., Papadimitratos, P., Hubaux, J.: Securing Vehicular Communications. IEEE Wireless Communications Magazine, Special Issue on Inter-Vehicular Communications 13(5), 8–15 (2006)

10. Dötzer, F.: Privacy Issues in Vehicular Ad Hoc Network. In: Danezis, G., Martin, D. (eds.) PET 2005. LNCS, vol. 3856, pp. 197–209. Springer, Heidelberg (2006)
11. Gerlach, M., Festag, A., Leinmüller, T., Goldacker, G., Harsch, C.: Security Architecture for Vehicular Communication. In: Workshop on Intelligent Transportation (2007)
12. Fonseca, E., Festag, A., Baldessari, R., Aguiar, R.: Support of Anonymity in VANETs - Putting Pseudonymity into Practice. In: IEEE Wireless Communications and Networking Conference (2007)
13. Schoch, E., Kargl, F., Leinmüller, T., Schlott, S., Papadimitratos, P.: Impact of pseudonym changes on geographic routing in vanets. In: European Workshop on Security in Ad-hoc and Sensor Networks, pp. 43–57 (2006)
14. Douceur, J.: The Sybil Attack. In: First International Workshop on Peer-to-Peer Systems, March 2002, pp. 251–260 (2002)
15. Raya, M., Hubaux, J.: Securing Vehicular Ad Hoc Networks. Journal of Computer Security, Special Issue on Security of Ad Hoc and Sensor Networks 15(1), 39–68 (2007)

Algebra for Capability Based Attack Correlation

Navneet Kumar Pandey, S. K. Gupta, and Shaveta Leekha

Indian Institute of Technology Delhi,
Delhi, India
{npandey,skg,mcs052943}@cse.iitd.ernet.in

Abstract. Most of the existing intrusion detection systems (IDS) often generate large numbers of alerts which contain numerous false positives and non relevant positives. Alert correlation techniques aim to aggregate and combine the outputs of single/multiple IDS to provide a concise and broad view of the security state of network. Capability based alert correlator uses notion of capability to correlate IDS alerts where capability is the abstract view of attack extracted from IDS alerts/alert. To make correlation process semantically correct and systematic, there is a strong need to identify the algebraic and set properties of capability. In this work, we identify the potential algebraic properties of capability in terms of operations, relations and inferences. These properties give better insight to understand the logical association between capabilities which will be helpful in making the system modular. This paper also presents variant of correlation algorithm by using these algebraic properties. To make these operations more realistic, existing capability model has been empowered by adding time-based notion which helps to avoid temporal ambiguity between capability instances. The comparison between basic model and proposed model is exhibited by demonstrating cases in which false positives have been removed that occurred due to temporal ambiguity.

Keywords: - *intrusion detection, capability model, attack scenario.*

1 Introduction

Since long information system and network security experts have made considerable efforts to protect secure systems from exponentially increasing threats, despite this the hackers tools now includes technology that conventional security tools and services cannot sustain. Even critical systems and networks equipped with highly sophisticated security techniques are vulnerable to blended and multi-stage attacks which use stealth and intelligence to strategically compromise a target, escaping detection and penetrating the defences [1].

The surveillance and security monitoring of the network infrastructure is mostly performed using Intrusion Detection Systems (IDSs). Event streams are used by IDS in two different ways, according to two different paradigms: anomaly detection and misuse detection. In anomaly detection systems ([2], [3], [4], [5], [6], [7]), historical data about a systems activity and/or specifications of the

J.A. Onieva et al. (Eds.): WISTP 2008, LNCS 5019, pp. 117–135, 2008.

intended behavior of users and applications are used to build a profile of the normal operation of the monitored system. The intrusion detection system then tries to identify patterns of activity that deviate from the defined profile. Misuse detection systems take a complementary approach ([8], [9], [10], [11], [12]). Misuse detection tools are equipped with a number of attacks descriptions (or signatures) that are matched against the stream of audit data and look for the evidence modeled attack. Misuse and anomaly detection systems have their own advantages and disadvantages [12].

Moreover, intrusion detection, audit and logging systems often provide sensory feedback data that cannot be effectively analyzed as they flag thousands of alerts which may overwhelm the analysts. Most of them are false positive and non relevant positives. Non relevant positives are alerts that correctly identify an attack, but the attack fails to meet its objective. Several alert correlation techniques have been proposed including approaches based on similarity between alert attributes, using pre-defined attack scenarios, pre/post-conditions of attacks, using multiple networks and auditing tools. Each technique has its own advantages and disadvantages, therefore none of the technique dominate the other [13]. For example similarity based approaches lack on finding attack step sequence, pre-defined attack scenario only work well for known scenarios, pre-post condition based approaches can detect new scenario but defining these conditions is itself error-prone and enumeration of these conditions is non trivial task whereas multiple information source based approaches suffer from sheer volume of data to process.

The require/provide model [14] used for alert correlation states that in a multistage intrusion comprising of a sequence of attacks, the early attacks acquire certain advantages, like information about the system under attack and the ability to perform actions on that system and use these advantages to support the later attacks that require them. Capability model [15] captures this notion of attacker capability and use it for logical alert correlation.

In this work we give algebraic property of capabilities. These properties give better understanding of capability characteristics. These characteristics help in designing the correlation process in a systematic and modular fashion. We group the identified algebraic properties into three classes i.e. operations, relations and inferences. Operations include join, split etc. which represent basic manipulation using one or more capability instance. Relations include overlapped, mutual exclusive, independent relations between capability instances. These relations help in identifying the preconditions to allow specific operations. As the whole system is based on require/provide model therefore to determine whether a capability satisfies a required capability, inferences are used. Inferences include comparable, resulting etc., which enumerate the possible inferences from different real life views. The paper also gives three derived version of the correlation process from basic correlation process using these algebraic properties. We have also enriched the basic capability model by adding time parameter in the definition of capability. This helps to remove temporal ambiguity between capability instances. The comparison between basic model and proposed model is exhibited by demonstrating cases.

In this work, we consider attack from single source to a single destination. However this can be easily generalized for distributed kind of attacks where multiple sources/multiple destinations are involved.

The remainder of this paper is structured as follows: Section 2 presents related work on alert correlation and capability model. Section 3 presents the proposed modified capability model. Section 4 presents the capability algebra. Section 5 provides a detail of the correlation algorithm and also shows the case study in which results have been improved. In the section 6, alternate way of correlation algorithm has been discussed with their benefits and pitfalls. Finally, Section 7 draws conclusions and outlines future work.

2   Related Work

In order to minimize false positives, the alert correlation techniques have been widely studied. Pouget et. al. [16] has classified these techniques into eight classes.(i) Rule based (ii) Scenario-based (iii) Uncertainty reasoning (iv)Time reasoning (v) State transition graphs (vi) Neural networks (vii) Bayesian belief networks (viii) Context reasoning. ([17] [18] [19] [20] [21] [22] [23] [24] [25])

To define logical relation between different attacks, Templeton and Levitt [14] proposed the require/provide model based on the system states. The proposed JIGSAW language for correlation uses simple predicates to define system state. However, they do not provide a systematic approach to develop predicates. Another similar approach given in Ning et. al. [26] also defines predicate for alert correlation. However definition of predicates used is ambiguous and also the paper does not give consistent way to develop it.

Our work is motivated by Jingmin et. al. [15] which uses capability model for attack correlation. This capability model uses capability as basic building block and used it for developing several algorithms in correlation based on alert abstraction and inference rules. Their work also shows that the approach is capable of handling missing attacks and is promising at alert fusion and correlation. However the paper does not discuss the algebraic operations and relation between capabilities.

In our work, we give a new definition of capability which is closer to the semantics of real life attacker capability and also avoid temporal ambiguity between definitions of capabilities. We also give several algebraic operations. The work also identified relations that exist between capabilities and derived inference rules to define logical association between two capabilities. These relations are helpful in understanding the capabilities properly and for defining the semantics of algebraic operations which in turn are used in correlation algorithm.

Li et. al [27] and Wijeskera et. al. [28] defined algebra for access control policies [29] which is related to capability. However the definition of capability is generalized as compared to access control and is not limited to a specific service e.g it applies for database, network, OS etc.

3   Capability Model

In the capability model, capability represents facilities and accesses that an attacker gains by making a connection. Capability describes the ability of an attacker during intrusion. An attacker can have many capabilities at a particular instance that may or may not belong to the same intrusion.

3.1   Capability Model

Let D be a set of network addresses, C be a set of credentials, A be a set of actions, SP be a set of pair of services and their property and $[t_1, t_2]$ is a time interval where t_1 and t_1 are constant time and $t_2 > t_1$.

Definition 1 (Capability). *Capability is a six-tuple capability = (source, destination, credential, action, (service, property), interval)*
where source $\in D$, destination $\in D$, credential $\in C$, action $\in A$, service $\in S$, property $\in P$, interval $\in [t_1, t_2]$ in which capability is valid. It may be noted that we have added the attribute interval to the definition of capability given by Jingmin et. al. [15].

Example 1. The capability (pushpa, dblab, user1, read,('/etc/passwd', content), *From* : $\langle 1997 - 07 - 16T19 : 20 : 30 + 01 : 00 \rangle$) means there is capability from host pushpa (source) to host dblab (destination) with credential user1 for read action of content of the file "/etc/passwd" in interval $[1997 - 07 - 16T19 : 20 : 30 + 01 : 00, \infty]$.

3.2   Attributes

In the modified capability model the definition of Source, Destination, Action and Credential are same as those given by [15]. As service and property attributes are tightly coupled, therefore they have been merged into a single attribute of two tuples. This modification helps us in defining operation more clearly. Interval is a new attribute added in the definition of capability. Table 1 shows description of each attribute along with examples.

3.3   Direct and Indirect Capability

After connection has been established from source (attacker host) to destination (victim host) by attacker, he/she may gain some privilege or knowledge. This type of capability will be called as direct capability. Implication of direct capability will be called indirect capability. For example if attacker succeed to make connection for reading a file in which mail and credit card passwords have been stored then direct capability is being able to read that file and indirect capability is being able to use mails and credit cards.

Table 1. Attributes of Capability

Attribute	Description	Example
Source	source address	IP:10.20.3.2, Ethernet:006097981E6B etc..
Destination	destination address	IP:10.20.3.2, Ethernet:006097981E6B etc..
Action	Actions that can be performed by an attacker	read, write, communicate etc.
Credential	Credential using which action can be done	root, system, user etc.
Service	Service used by connection	IIS 3.0, \etc\passwd etc.
Property	Property of service	version, content etc.
Interval	Capability Time interval during	From:tstamp, Between:tstamp+tt, at:tstamp etc.. Where tstamp time-stamp and tt is length

3.4   Significance of Time Parameter

Time parameter which is denoted by the interval is a crucial parameter in reducing the false belief that each capability will last forever during correlation. Some capabilities especially indirect capabilities that depend on service running in target host and may only be valid under certain conditions. It is not necessary that these conditions will always be present in the network or system for example in case where service is scheduled to run for a specific duration. Therefore, it is clear that these conditions are bound to the validity of a session for capability and cannot be assumed that once gained by attacker they will always be with him. Ambiguity due to the assumption that capability once gained will always exist is called temporal ambiguity.

Time interval is represented by predicate *between* : $[t1, t2]$ which shows that capability will exist from timestamp t1 to timestamp t2. It is also true that some capabilities (e.g. of knowledge type) once gained will always exist with the attacker. To denote it *from* predicate is used i.e. *From* : $t1$. For example if attacker gained that target machine running on Solaris OS at time t1 then interval of this capability is *From* : $t1$.

There are various sources of information that may help in specifying the closed time interval of the capability e.g. host integrity checker, HIDS etc... From these sources it can be identified that capability gained earlier is no longer valid. Other sources may be administrator knowledge especially when some service is allowed for a limited period. For example there is one connection having capability to execute a program in host at time t_1 and later at time t_2(where $t_2 > t_1$) service has been blocked . In this case former established connection will not have same capability as in earlier.

Timestamp can be taken in different format. In this model following format of timestamp has been used:-

$$YYYY\text{-}MM\text{-}DDThh{:}mm{:}ss.sTZD$$

For example 2007-07-16T19:20:30.45+01:00

where YYYY = four-digit year, MM = two-digit month (01=January, etc.), DD = two-digit day of month (01 through 31), hh = two digits of hour (00 through 23) (am/pm NOT allowed), mm = two digits of minute (00 through 59), ss = two digits of second (00 through 59), s = one or more digits representing a decimal fraction of a second, TZD = time zone designator (Z or +hh:mm or -hh:mm).

3.5   Correlation Process

The correlation process is based on the require/provide model in which capabilities gained from the previous attacks are used to satisfy the prerequisite of subsequent attacks. The model has following components.

H-alert. An H-alert is a three tuple (require, provide, raw) and represents transformed object of alert in terms of capability, where
   Require: - It is a set of capabilities that are required for alert to be a true attack.
   Provide: - It is a set of gained capabilities after an alert has been generated. Most IDS generate two kinds of alerts for each attack step, one for incoming traffic in victim host and the other for outgoing traffic from victim host. Alerts that have been generated for incoming traffic may be either successful or failure. This information is available in outgoing traffic. Attacker may even gain capability in failed attack therefore provide set contain those capabilities which have been gained by either successful attack or failure attack whichever is applied.
   Raw: - Raw contains other information available in alert message such as time of alert generated, traffic direction etc.

M-Attack. An M attack is a three tuple $(haset, capset, tmpstmp)$ which is a collection of correlated alerts where $haset$ is a set of alerts $(h-alerts)$, $capset$ is a set of capabilities provided by h-alerts in $haset$ and $tmpstmp$ is the timestamp of last correlated alert which can be considered as $timestamp$ of M.
   Capabilities are tagged to be considered as mandatory and optional (can be ignored while correlation in some conditions) in the $capset$.
   In other words, M-attack represents the set of correlated alerts and correlation process correlate the newly generated alert (H-alert) with these M-attack/M-attacks. Overall correlation algorithm has been explained in section 5.1.

4   Capability Algebra

To modularize the whole correlation process, it is necessary to analyze the properties and characteristics of the capability model. By identifying the algebraic properties of capabilities, capability extraction from IDS signatures can be made automatic. It gives better insight and puts clarity and separation between definitions of capabilities. This also helps to determine the level of granularity in defining the capability. Capability algebra can be divided into three groups i.e. operations, relations and inferences. These are described in the following section.

For comparing two capabilities, it is required to determine the relations between two capabilities, their inferences and relevant operations. It may be noted that attributes of capabilities form a hierarchy. We identify following operations, relations and inferences base on this hierarchy.

4.1 Operations

Operations represent manipulations in the capabilities required in the correlation process. There are three kinds of operations identified for the correlation process.

Join. Join operation merges two capabilities in presence of a join condition (see Algorithm 1). Two capabilities can be joined if both capabilities belong to the same source and destination. Also other attributes should be same except an attribute based on which join operation will be performed. For example capability C_1 (srcS, dstD, daemon, block, (ftp process, port 80), from:2008-07-16T19:20:30.45+01:00) is join capability of C_2 (srcS, dstD, daemon, block, (ftp process, port 80), between:[2008-07-16T20:10:30.00+01:00, 2008-07-16T21:00:00 .00+01:00]) and C_3 (srcS, dstD, daemon, block, (ftp process, port 80), from:2008-07-16T19:20:30.45+01:00).

Algorithm 1. Joining two capabilities

Require: Two capabilities C_1 and C_2
Ensure: Resultant capability C_3 if C_1 and C_2 can be joined else $NULL$.
  Let $S = (cred, action, (service, property))$
  procedure JOIN(C_1,C_2)
    if $C_1.src = C_2.src$ and $C_1.dst = C_2.dst$ **then**
      if $\forall A_i \in S$ s.t. $C_1.A_i = C_2.A_i$ **then return** C_1
      else if $\exists$ an attribute $A_k \in S$ s.t. $C_1.A_k \neq C_2.A_k$ and $\forall A_i \in S - A_k$,
        $C_1.A_i = C_2.A_i$ **then return** C_3 with $C_3.A_k = C_1.A_k \cup C_2.A_k$
      else if $C_1.interval$ and $C_2.interval$ overlaps and other attributes are same
        then return C_3 with $C_3.interval = C_1.interval \cup C_2.interval$
      end if
    end if
    return NULL
  end procedure

Join operation reduces the redundancy which in turn minimizes the number of comparisons (while finding inferences) between h-alert require set and M-attacks Capset (see section 5) during correlation process.

Split. Split breaks a capability into two capabilities based on the given attribute and its value. For example (srcS, dstD, userU, modify, (file, content), from:t1) can be split in (srcS, dstD, userU, append, (file, content), from:t1) and (srcS, dstD, userU, delete, (file, content), from:t1). It may be noted that split is semantically inverse of join operation. Split can be performed on the attributes (a) Action

(b) Credential (c) Property (d) Time, if their value is composite[1]. After split resultant capabilities would have same values of src(source), dst (destination) and service, however no split will be done on the basis of these attributes.

Split is a special case of Reduce (defined in section 4.1) where one capability C when split in two capabilities C_1 and C_2 then by joining C_1 and C_2 we can form C again which may not be the case in Reduce. In other words, split is lossless reduction (see Algorithm 2).

Algorithm 2. Split a capability into two capabilities for given attribute

---

Require: Capability C, Attribute A and value of attribute v
Ensure: Resultant capability C_1 and C_2 if C can be split else C
    procedure SPLIT(C,A,v)
        if $C.A$ is not *composite*[1] **then return** C
        else $C_1.A = v$, $C_2.A = reduce(C, A, v)$, $\forall A_i \in S - A$ set $C_1.A_i = C_2.A_i = C.A_i$
            where S=(src, dst, cred, action, (service, property), interval)
            return C_1 and C_2
        end if
    end procedure

Reduce. The Reduce operation weakens a capability by reducing strength of any of its attribute. For example capability (srcS, dstD, root, modify, (program, code), from:t1) can be reduced to (srcS, dstD, userU, modify, (file, content), from:t1). Difference between *split* (Algorithm 2) and *reduce* (Algorithm 3) is that *split* operation always gives two capabilities whereas in the case of *reduce* it is not mandatory that reduced part will be a capability.

Algorithm 3. Reducing a capability

Require: Capability C, Attribute A (must be composite) and v is value of A
Ensure: Reduced capability C_d
    Let $S = (src, dst, cred, action, (service, property), interval)$
    procedure REDUCE(C,A,v)
        Create a new capability C_d with $C_d.A = C_d.A - v$,
        $\forall A_i \in S - A$ set $C_d.A_i = C.A_i$, **return** C_d
    end procedure

Subtract. The Subtract operation takes two capabilities C_1 and C_2 and returns C_3 which is deduction of capability C_2 from C_1. For example (srcS, dstD, userU, send, (IIS, Ftp), from:t1) is result of subtraction of (srcS, dstD, userU, receive, (IIS, Ftp), from:t1) from (srcS, dstD, userU, communicate, (IIS, Ftp), from:t1).

Subtract is similar to reduce in which minuend Capability is reduced by subtrahend capability. For the substraction it is necessary that both capabilities

---

[1] Attribute A is composite if it contains multiple values or a value that can be divided into distinct components for eg. RW action can be split into R and W actions.

Algorithm 4. Capability Subtraction

Require: Capabilities C_1 and C_2
Ensure: Resultant capability C_s
  procedure SUBTRACT(C_1,C_2)
    if $C_1.src=C_2.src$ and $C_1.dst=C_2.dst$ **then**
      if $\exists$ an attribute $A \in S$ s.t. $C_1.A \neq C_2.A$ and $\forall B \in S - A$, $C_1.B = C_2.B$
        where S=(action, (service, property), interval) **then**
          $C_s=$ Reduce(C_1,A,$C_2.A$)
      else $C_s = C_1$.
      end if return C_s
    end if
  end procedure

---

have same source and destination and only one attribute is different among the rest (see Algorithm 4).

4.2 Relations

A relation represents a logical association between two or more capabilities. Following three types of relations are identified for the correlation process.

Overlap. Two capabilities overlap if there exists a common capability between them (see Algorithm 5). For example capabilities (SLab, Dlab, RW, (/home/user1, content), user1, from:t1) and (SLab, Dlab, WX, (/home/user1, content), user1, from:t1) overlap because the capability (SLab, Dlab, W, (/home/user1, content), user1, from:t1) is common in both. If any of the following attributes are common in two capabilities, then there is overlapping: (a) Interval (b) Credential (c) Action and property of service.

Algorithm 5. Test two capabilities whether they are overlap

Require: Two capabilities C_1 and C_2
Ensure: *true* or *false*
  Let $S = (src, dst, cred, action, (service, property), interval)$
  procedure OVERLAP(C_1,C_2)
    if ($C_1.interval$ and $C_2.interval$ overlaps) **and** $\forall A_i \in S - \{interval\}$
      s.t. $C_1.A_i = C_2.A_i$ **then return** *true*
    else if $\exists$ a credential $cred_k$ s.t. $cred_k \in C_1.cred \cap C_2.cred$ **and** $\forall A_i \in S - \{cred\}$
      s.t. $C_1.A_i = C_2.A_i$ **then return** *true*
    else if $\exists$ an action act_k s.t. $act_k \in C_1.action \cap C_2.action$ **and** $\forall A_i \in S - \{action\}$
      s.t. $C_1.A_i = C_2.A_i$ **then return** *true*
    else if $\exists$ a property p s.t. $p \in C_1.property \cap C_2.property$ of the same service **and**
      $\forall A_i \in S - \{(serivce, property)\}$ s.t. $C_1.A_i = C_2.A_i$ **then return** *true*
    else return *false*
    end if
  end procedure

Independent. Two capabilities are independent if they cannot be joined (see Algorithm 6).In other words, two capabilities are called independent if either both have different source/destination or have different values of more than one attributes among the rest of attributes. For example capabilities (SLab, Dlab, W, /home/user1, content, user1, from:t1) and (SLab, Dlab, X, httpd, (Apache 3.2, apacU), from:t1) independent.

Algorithm 6. Test two capabilities whether they are Independent

---

Require: Two capabilities C_1 and C_2
Ensure: *true* or *false*
  procedure INDEPENDENT(C_1,C_2)
     if join(C_1,C_2) is NULL **then return** *true*
     else return *false*
     end if
  end procedure

Mutual Exclusive. Two capabilities are mutually exclusive if their corresponding attribute's value cannot coexist (see Algorithm 7). Mutually exclusive capabilities are less likely to belong to the same attack. This information helps in reducing false correlation. For example capabilities (SLab, Dlab, R,(/etc/passwd, content), user1, from:t1) and (SLab, Dlab, X, IIS, Ver4.0, user1, from:t1) are mutually exclusive.

Algorithm 7. Test two capabilities whether they are Mutual Exclusive

---

Require: Two capabilities C_1 and C_2
Ensure: *true* or *false*
  procedure MUTUAL-EXCLUSIVE(C_1,C_2)
     if $\exists$ an attribute A s.t. conflict($C_1.A$,$C_2.A$) is *true* **then return** *true*
     else return *false*
     end if
  end procedure

The conflict set used in algorithm 7 is a knowledge base having pair of attributes that cannot coexist e.g. service of windows and Linux cannot exist simultaneously in the same IP.

4.3   Inferences

Inference means causal relationship involved in process of deriving result or making a logical judgment on the basis of known evidence. Inferences identified here are used in comparing capabilities of require set of h-alert with capabilities in M-attack's capability set based on require/provide model during correlation process. Almost all inferences given in this section are same as given in [15].

Comparable Inference. Comparable inference denotes semantic comparability of two capabilities. Two capabilities can be compared only if they hold same type of service and property while other attributes must be same. This inference will be used to correlate two capabilities to construct attack scenario. Capabilities can be correlated only if required capability can be satisfied with some of the capability of M-attack set by comparable inference (see Algorithm 8).

Algorithm 8. Test whether C_1 and C_2 can be compare directly

---

Require: Two capabilities C_1 and C_2
Ensure: *true* or *false*
    procedure COMPARABLE(C_1,C_2)
        if $\forall A_i \in \{$ src, dst, cred, action $\}$, $C_1.A_i = C_2.A_i$, with overlapped time interval
            and both have same type of service and property **then return** *true*
        else return *false*
        end if
    end procedure

Service and property belong to the same type when services belong to same category as given in [15].

Resulting Inference. In many cases logical relations between capabilities cannot be represented by comparable inference due to strict conditions. One capability is the resulting inference of other if it gives the other capability on its execution. These inferences are nothing but a single step of whole correlation process and are used in making attack scenario through multi step correlations (see Algorithm 9).

Algorithm 9. Test whether C_2 is resulting inferable from C_1

Require: Two capabilities C_1 and C_2
Ensure: *true* or *false*
    procedure RESULTING_INFERABLE(C_1,C_2)
        if exercise of C_1 logically derive C_2 **then return** *true*
        else return *false*
        end if
    end procedure

Administrator knowledge, topology of network are some of the major information sources to identify the capabilities which can be logically derived by exercising a capability.

Other Inferences. Several other inferences are also possible along with the given inferences. For example compromise inference and external inference as given by Jingmin et. al. [15].Through compromise inference one capability can be inferred from other capability for compromising the destination machine (executing arbitrary program).Capability C_1 can be externally inferred from capability C_2 if C_2 is the capability to execute arbitrary program on destination machine which is the source of C_1.

5   Correlating Alert Using Modified Capability Model

5.1   Correlation Algorithm

Correlation algorithm correlates new h-alert (created from alert generated by IDS) with the existing M-attacks. Initially there is a set of M-attacks M. Whenever a new alert comes, then it is abstracted into an h-alert. Correlation algorithm searches minimal and ordered subset of M-attacks from M such that all the require capabilities of h-alert are satisfied by the capabilities of a subset of M-attacks. Then the algorithm combines h-alert with the identified subset of M-attacks in a single M-attack. This M-attack contains all capabilities of selected M-attacks along with the h-alert's provide capabilities. This new M-attack replaces the subset of M-attacks. The whole correlation process is presented in Algorithm 10. Algorithm 11 shows the search procedure of M-attacks that satisfy the required capabilities of newly generated h-alert.

5.2   Case Study

We have extended the existing capability model by adding a new attribute i.e. time. The modified capability improves the correlation by reducing the cases of false correlation and by increasing correlation strength. Some of the major cases are as follows.

Case1:   We have a require capability C_1 (srcX, dstX, credX, $\{RW\}$,(/home/ user1, content), intvX) of a newly generated h-alert and two M-attacks M_1 and M_2 in M-attack set having capabilities C_3 (srcX, dstX, credX, $\{R\}$,(/home/ user1, content), intvX), C_3(srcX, dstX, credX, $\{W\}$,(/home/user1, content), intvX) in their *capset* respectively. Using former approach capability C_1 cannot be correlated with either of M-attacks (M_1 or M_2) capability because the action attribute of C_1 cannot directly be compared with that of C_2 or C_3. Therefore, the former approach is unable to correlate it. But in the modified approach when C_1 and C_2 will be correlated, C_1 will reduce to (srcX, dstX, credX, W,(/home/user1, content), intvX) and it is directly correlated with C_3 i.e. C_1 is correlated by $M_1 \cup M_2$. Consequently, the enhanced model is able to detect these kinds of true correlations that would have gone undetected in earlier approach. These kind of cases have been handled in the modified approach because of flexibility by defined operations.

Case2:   Consider another case where the require set of an incoming h-alert is satisfied by the *capset* of M-attacks M_1, M_2 and there exists a capability in M_1 which is mutually exclusive of other capability that belongs to M_2. In

---

Algorithm 10. Correlate a new h-alert which is an abstract form of recently came alert with M-attacks

---

Require: h-alert h_1 and set of M-attacks $M=\{M_1, \cdots M_n\}$
Ensure: a new M-attacks set M'$=\{M_1, \dots M_k\}$
   procedure CORRELATION ALGORITHM(h_1,M)
      Find a minimal and ordered subset M^k of set M (as given in Algorithm 11)
      such that h1.requires is satisfied by capabilities in M-attacks of set M^k
      if $M^k \neq \phi$ **then** Make new M-attack M_{new} as
          $M_{new}.capset = C_M \cup h_1.provide$ where $C_M = \bigcup_i M_i^k.capset$ and $M_i^k \in M^k$,
          $M_{new}.haset = h_M \cup h_1$ where $h_M = \bigcup_i M_i^k.haset$ and $M_i^k \in M^k$
          and replace all M-attacks in M^k by M_{new}
      else Make new M_{new} as $M_{new}.capset = h_1.provide$ and $M_{new}.haset = h_1$
      end if
      $M_{new}.timestamp$ equal to the timestamp of newly correlated alert.
   end procedure

---

this case M_1 and M_2 actually have no correlation. But the former approach could correlate these kind of capabilities. Whereas, in the proposed model such capabilities are not correlated because they are mutually exclusive and logically donot belong to the same attack.

For example a capability C_1 (eth0:12ffdd3453, eth0:12ffee1234, credX, $\{RW\}$, (/home/user1, content), intvX) belongs to M_1 and other capability C_2 (srcX, dstX, credX, $\{RW\}$, (IIS, content), intvX) belongs to M_2. Administrator knowledge, services running in the network, topology of network are the major sources of domain knowledge in identifying the mutual exclusive capabilities discussed in section 4.2.

Case3: Modified process also handles the correlation conflicts that arise due to temporal ambiguity as explained in section 3.4. For example, suppose attacker has a capability to read and write in a host H, then attacker can also read and write the mail of a user whenever he opens his mail account on that machine i.e. attacker will have the capability of reading/writing mail from a particular user account only for the duration in which the user is logged in. However in this case, there is no upper limit of interval for reading/writing other files. To avoid this ambiguity time attribute has been added with every capability.

Apart from the cases discussed above, there are several other cases where proposed model helps in making overall process efficient. For example *Join* operation helps in reducing the redundancy which in turn saves the number of comparisons while correlations. Suppose there are two capabilities C_1 (srcX, dstX, credX, $\{RW\}$, (/, content), intvX) and C_2 (srcX, dstX, credX, $\{RW\}$, (/home/, content), intvX) then we can join these two into one capability as they are forming contain-ship relation. Therefore it is clear that if two capabilities of M-attack's cap set are joined then further correlation needs only one comparison instead of two. Overlapped and independent relations help in defining join condition accurately to test unambiguously that two capabilities can be joined or not.

Algorithm 11. Find a minimal and ordered subset M^k of set M

Require: h-alert h_1 and set of M-attacks $M=\{M_1, \cdots M_n\}$
Ensure: M-attacks set M'
1: **procedure** FIND_MIN_ORDERED_SUBSET(h_1,M)
2:     Order all M-attacks based on decreasing Timestamp and let cap_{req_set} is set of capabilities in $h_1.require$, $CapM_{sat} = \phi$ and $M_{result} = \phi$
3:         **for all** M-attacks $M_i \in \{M_1, \cdots M_n\}$ **do** find subset $cap_{satisfied} \subseteq cap_{req\_set}$ $inferable$ (see section 4.3) from $M_i.capset$, $CapM_{sat} = CapM_{sat} \cup cap_{satisfied}$
4:             **if** $CapM_{sat} = h_{req\_set}$ **then** $M_{result} = M_{result} \cup M_i$ and **return** M_{result}
5:             **else if** $cap_{satisfied} \neq \phi$ **then**
6:                 $cap_{req\_set} = cap_{req\_set}$-$cap_{satisfied}$ and $M_{result} = M_{result} \cup M_i$
7:             **else**
8:                 find $cap_{sub} \in cap_{req\_set}$ that can be obtained from $M_i.capset$ by subtract.
9:                 **if** $cap_{sub} \neq \phi$ **then**
10:                     $cap_{req\_set} = cap_{req\_set} - cap_{sub}$, and $M_{result} = M_{result} \cup M_i$
11:                 **end if**
12:             **end if**
13:         **end for**
14:         **return** ϕ
15: **end procedure**

---

6   Discussion and Other Issues

In this section other possible ways of correlation process are discussed. It is clear that join algorithm has significant impact in minimizing the number of comparisons in correlation because it combines the capabilities in M-attacks's capset. However join itself is costlier operation in terms of time as described below. Following are the alternate methods of doing correlation using various combinations of join and split.

6.1   Alternate Method 1

In this method after the correlation, algorithm 12 joins capabilities within each M-attack i.e. within each M-attack if two or more capabilities can be joined then they are joined to minimize the number of capabilities in capset and removes the redundancy if it is there. The minimal set search algorithm is same as algorithm 11.

It may be noted that the method minimizes the number of comparisons while searching for the minimal set of M-attacks because of lesser number of capabilities in each M-attack's capset.

However join operation is a costlier operation. For example in a M-attack's capset if there are n capabilities then join operation is called for every pair of subset of capabilities which is exponential because the join operation need to be called recursively until no more joins are possible.

Algorithm 12. (Alternate Method 1) Correlate a new h-alert which is an abstract form of recently came alert with M-attacks

Require: h-alert h_1 and set of M-attacks $M=\{M_1, \cdots M_n\}$
Ensure: a new M-attacks set M'=$\{M_1, \ldots M_k\}$
  procedure CORRELATION ALGORITHMII(h_1,M)
    Find a minimal and ordered subset M^k of set M such that h1.requires is
    satisfied by capabilities in M-attacks of set M^k using algorithm 11
    if $M^k \neq \phi$ **then** Make new M_{new} as
      $M_{new}.capset = C_M \cup h_1.provide$ where $C_M = \bigcup_i M_i^k.capset$ and $M_i^k \in M^k$,
      $M_{new}.haset = h_M \cup h_1$ where $h_M = \bigcup_i M_i^k.haset$
      and $M_i^k \in M^k$ and replace all M-attacks in M^k by M_{new}
    else Make new M_{new} as $M_{new}.haset = h_1.provide$ and $M_{new}.haset = h_1$
    end if
    for all pair of capabilities (C_i, C_j) in $M_{new}.capset$ **do** C_k=join(C_i, C_j)
      if $C_k \neq NULL$ **then** replace C_i and C_j by C_k in $M_{new}.capset$
      end if
    end for
    M_{new}.timestamp equal to the timestamp of newly correlated alert.
  end procedure

6.2 Alternate Method 2

In this method capabilities in new h-alert's require set are split into minimal granularity based on their composite attributes.

In this case, we do not use join operation for correlation as it is costly. By using split operation, the granularity of each attribute of every capability will become one. Consequently, this will make the comparisons easier. Also we do not need the subtract operation as all capabilities are in their minimal reduced form. Minimal set search algorithm is same as algorithm 11 except the subtract operation in steps 8,9 and 10.

However in some cases we may end up in split where it may not be required. For example capability containing action RW has been split into two capabilities with action R and W in M-attack's capset. A new required capability with same RW action comes, then we split it into R and W, which require two comparisons. Indirectly we may be increasing the comparisons unintentionally as number of capabilities in the capset of M-attack have increased in some cases.

6.3 Alternate Method 3

This method is a combination of Alternate Method 1 and Alternate Method 2 which splits the capabilities of h-alert's require set into minimal granules and after correlation, joins the capabilities in the newly formed M-attacks's capset which can be joined.

The method wipes out pitfall of pervious alternate methods as split has been used initially to simplify the comparisons and later on join has been used in each M-attack's capset to minimize the number of capabilities which consequently

Algorithm 13. (Alternate Method 2) Correlate a new h-alert which is an abstract form of recently came alert with M-attacks set

Require: h-alert h_1 and set of M-attacks $M=\{M_1, \cdots M_n\}$
Ensure: a new M-attacks set M'=$\{M_1, \ldots M_k\}$
  Let $S = \{cred, action, \{service, property\}, interval\}$
  procedure CORRELATION ALGORITHMIII(h_1,M)
      for all capabilites $C_i \in h_1^{\sim}.require$ **do**
        for all attributes $A \in S$ **do**
          if A is composite **then** split C_i into minimal granularity based on A
          end if
        end for
      end for
      Find a minimal and ordered subset M^k of set M such that h1.requires is satisfied by capabilities in M-attacks of set M^k using algorithm 11
      if $M^k \neq \phi$ **then** Make new M_{new} as
        $M_{new}.capset = C_M \cup h_1.provide$ where $C_M = \bigcup_i M_i^k.capset$ and $M_i^k \in M^k$,
        $M_{new}.haset = h_M \cup h_1$ where $h_M = \bigcup_i M_i^k.haset$ and $M_i^k \in M^k$
        and replace all M-attacks in M^k by M_{new}
      else Make new M_{new} as $M_{new}.capset = h_1.provide$ and $M_{new}.haset = h_1$
      end if
      M_{new}.timestamp equal to the timestamp of newly correlated alert.
  end procedure

minimizes the number of comparisons . However this method is more costly than previous in time complexity.

7   Conclusion

In this work we have defined time parameter and shown its impact in reducing false correlation. We have also identified and defined relations between capabilities, operations on capability and derived Inference rules along with their semantic that have been used in correlation process. The framework is made systematic, consistent and defined properly with algorithms. Comparison between the previous model and the proposed model is exhibited by demonstrating cases where the correlated alerts were not captured by the old model, but are taken care in our proposed model.

By making the correlation process modular we have simplified the whole correlation process. This makes system more understandable for even non security expert. This approach helps in facilitating the process flexibility and easy enhancement. With this systematic model, the system can be automated and adaptive to optimizations.

Part of the future work will be to optimize algorithms and to achieve better performance. One possibility would be to optimize the algorithm of join operation and to use that in given alternate correlation algorithm (in section 6). This would help in making whole system real time with low false rate.

Algorithm 14. (Alternate Method 3) Correlate a new h-alert which is an abstract form of recently came alert with M-attacks set

---

Require: h-alert h_1 and set of M-attacks $M=\{M_1, \cdots M_n\}$
Ensure: a new M-attacks set M'=$\{M_1, \ldots M_k\}$
  Let $S = \{cred, action, \{service, property\}, interval\}$
  procedure CORRELATION ALGORITHMIV(h_1,M)
    for all capabilites $C_i \in h_1.require$ **do**
      for all attribute $A \in S$ **do**
        if A is composite **then** split C_i into maximum granularity based on A
        end if
      end for
    end for
    Find a minimal and ordered subset M^k of set M such that h1.requires is satisfied by capabilities in M-attacks of set M^k using algorithm 11
    if $M^k \neq \phi$ **then** Make new M_{new} as
      $M_{new}.capset = C_M \cup h_1.provide$ where $C_M = \bigcup_i M_i^k.capset$ and $M_i^k \in M^k$,
      $M_{new}.haset = h_M \cup h_1$ where $h_M = \bigcup_i M_i^k.haset$ and $M_i^k \in M^k$
      and replace all M-attacks in M^k by M_{new}
    else Make new M_{new} as $M_{new}.capset = h_1.provide$ and $M_{new}.haset = h_1$
    end if
    for all pair of capabilities (C_i,C_j) in $M_{new}.capset$ **do** C_k=join(C_i,C_j)
      if $C_k \neq NULL$ **then** replace C_i and C_j by C_k in $M_{new}.capset$
      end if
    end for
    M_{new}.timestamp equal to the timestamp of newly correlated alert.
  end procedure

Another future work will be to model the defence capability of security personnel. This defence capability will help the administrator in identifying his position against the attacker's capability. There is also scope in the future work to develop language for whole framework.

Acknowledgment

This project was supported by a grant from Dept of Information Technology, Govt of India. We thank the database and security group members at IIT Delhi for their reviews and valuable comments to improve this paper.

References

1. Dawkins, J., Hale, J.: A systematic approach to multi-stage network attack analysis. In: Information Assurance Workshop, 2004. Proceedings. Second IEEE International, April 8-9, 2004, pp. 48–56 (2004)
2. Denning, D.E.: An intrusion-detection model. IEEE Trans. Softw. Eng. 13(2), 222–232 (1987)

3. Gosh, A.K., Wanken, J., Charron, F.: Detecting anomalous and unknown intrusions against programs. In: ACSAC 1998: Proceedings of the 14th Annual Computer Security Applications Conference, Washington, DC, USA, p. 259. IEEE Computer Society, Los Alamitos (1998)
4. Javits, V.: The NIDES statistical component: Description and justification (March 1993), http://www.csl.sri.com/papers/statreport
5. Ko, C., Ruschitzka, M., Levitt, K.: Execution monitoring of security-critical programs in distributed systems: a specification-based approach. In: SP 1997: Proceedings of the 1997 IEEE Symposium on Security and Privacy, Washington, DC, USA, p. 175. IEEE Computer Society, Los Alamitos (1997)
6. Kruegel, C., Vigna, G.: Anomaly detection of web-based attacks. In: CCS 2003: Proceedings of the 10th ACM conference on Computer and communications security, pp. 251–261. ACM, New York, NY, USA (2003)
7. Warrender, C., Forrest, S., Pearlmutter, B.A.: Detecting intrusions using system calls: alternative data models. Security and Privacy, 1999. In: Proceedings of the 1999 IEEE Symposium on Security and Privacy, 133–145 (1999)
8. Paxson, V.: Bro: a system for detecting network intruders in real-time. Computer Networks 31(23-24), 2435–2463 (1999)
9. Neumann, P.G., Porras, P.A.: Experience with emerald to date. In: Proceedings of the Workshop on Intrusion Detection and Network Monitoring, Berkeley, CA, USA, pp. 73–80. USENIX Association (1999)
10. Roesch, M.: Snort - lightweight intrusion detection for networks. In: LISA 1999: Proceedings of the 13th USENIX conference on System administration, Berkeley, CA, USA, pp. 229–238. USENIX Association (1999)
11. Vigna, G., Kemmerer, R.A.: Netstat: a network-based intrusion detection system. J. Comput. Secur. 7(1), 37–71 (1999)
12. Eckmann, S.T., Vigna, G., Kemmerer, R.A.: Statl: an attack language for state-based intrusion detection. J. Comput. Secur. 10(1-2), 71–103 (2002)
13. Xu, D., Ning, P.: Alert correlation through triggering events and common resources. In: Yew, P.-C., Xue, J. (eds.) ACSAC 2004. LNCS, vol. 3189, pp. 360–369. Springer, Heidelberg (2004)
14. Templeton, S.J., Levitt, K.: A requires/provides model for computer attacks. In: NSPW 2000: Proceedings of the 2000 workshop on New security paradigms, pp. 31–38. ACM, New York (2000)
15. Zhou, J., Heckman, M., Reynolds, B., Carlson, A., Bishop, M.: Modeling network intrusion detection alerts for correlation. ACM Trans. Inf. System Secur. 10(1), 4 (2007)
16. Pouget, F., Dacier, M.: Alert correlation: Review of the state of the art. Technical Report EURECOM+1271, Institut Eurecom, France (December 2003)
17. Manganaris, S., Christensen, M., Zerkle, D., Hermiz, K.: A data mining analysis of rtid alarms. Computer Networks 34(4), 571–577 (2000)
18. Michel, C., Mé, L.: Adele: an attack description language for knowledge-based intrustion detection. In: Sec 2001: Proceedings of the 16th international conference on Information security: Trusted information, pp. 353–368 (2001)
19. Cuppens, F., Miège, A.: Alert correlation in a cooperative intrusion detection framework. In: SP 2002: Proceedings of the 2002 IEEE Symposium on Security and Privacy, Washington, DC, USA, p. 202. IEEE Computer Society, Los Alamitos (2002)
20. Siraj, A., Vaughn, R.B.: Alert correlation with abstract incident modeling in a multi-sensor environment. IJCSNS International Journal of Computer Science and Network Security 7(8), 8–19 (2007)

21. Morin, B., Mé, L., Debar, H.: Correlation of Intrusion Symptoms: An Application of Chronicles. In: Vigna, G., Krügel, C., Jonsson, E. (eds.) RAID 2003. LNCS, vol. 2820, pp. 94–112. Springer, Heidelberg (2003)
22. Vigna, G., Valeur, F., Kemmerer, R.A.: Designing and implementing a family of intrusion detection systems. In: ESEC/FSE-11: Proceedings of the 9th European software engineering conference held jointly with 11th ACM SIGSOFT international symposium on Foundations of software engineering, pp. 88–97. ACM, New York (2003)
23. Yang, D., Chen, G., Wang, H., Liao, X.: Learning vector quantization neural network method for network intrusion detection. Wuhan University Journal of Natural Sciences 12(1), 147–150 (2007)
24. Mehdi, M., Zair, S., Anou, A., Bensebti, M.: A bayesian networks in intrusion detection systems. Journal of Computer Science 3(5), 259–265 (2007)
25. Morin, B., Mé, L., Debar, H., Ducassé, M.: M2d2: A formal data model for ids alert correlation. In: Wespi, A., Vigna, G., Deri, L. (eds.) RAID 2002. LNCS, vol. 2516, pp. 115–137. Springer, Heidelberg (2002)
26. Ning, P., Cui, Y., Reeves, D.S., Xu, D.: Techniques and tools for analyzing intrusion alerts. ACM Trans. Inf. Syst. Secur. 7(2), 274–318 (2004)
27. Li, N., Wang, Q.: Beyond separation of duty: an algebra for specifying high-level security policies. In: CCS 2006: Proceedings of the 13th ACM conference on Computer and communications security, pp. 356–369. ACM, New York (2006)
28. Wijesekera, D., Jajodia, S.: A propositional policy algebra for access control. ACM Trans. Inf. Syst. Secur. 6(2), 286–325 (2003)
29. Bonatti, P., De Capitani di Vimercati, S., Samarati, P.: An algebra for composing access control policies. ACM Trans. Inf. Syst. Secur. 5(1), 1–35 (2002)

On the BRIP Algorithms Security for RSA

Frédéric Amiel[1] and Benoit Feix[2]

[1] AMESYS,
1030, Avenue Guillibert de la Lauzire,
13794 Aix-en-Provence, Cedex 3, France
f.amiel@amesys.fr
[2] INSIDE CONTACTLESS
41 Parc Club du Golf
13856 Aix-en-Provence, Cedex 3, France
bfeix@insidefr.com

Abstract. Power Analysis has been intensively studied since the first publications in 1996 and many related attacks on naive implementations have been proposed. Nowadays algorithms in tamper resistant devices are protected by different countermeasures most often based on data randomization such as the BRIP algorithm on ECC and its RSA derivative. However not all of them are really secure or in the best case proven to be secure. In 2005, Yen, Lien, Moon and Ha introduced theoretical power attacks on some classical and BRIP exponentiation implementations, characterized by the use of a chosen input message value ± 1. The first part of our article presents an optimized implementation for BRIP that takes advantage of the Montgomery modular arithmetic to speed up the mask inversion operation. An extension of the Yen *et al.* attack, based on collision detection through power analysis, is also presented. Based on this analysis we give security advice on this countermeasure implementation and determine the minimal random length to reach an appropriate level of security.

Keywords: Power analysis, collision attacks, RSA, BRIP, modular multiplication and exponentiation.

1 Introduction

Asymmetric cryptography was introduced by Diffie and Hellman [DH76] in 1976. The most widely used algorithms today are: RSA [RSA78] invented in 1978 by Rivest, Shamir, and Adleman, and elliptic curve cryptosystems (ECC) independently introduced by Koblitz [Kob87] and Miller [Mil86].

Compared with symmetric cryptography, public key algorithms are computationally very intensive. In practice long integer arithmetic is most often handled by specific coprocessors designed for efficient computation in $GF(p)$. This is the case for embedded solutions with strict power consumption and/or timing constraints.

Initially smart cards were considered inherently tamper resistant because any private data was embedded and thus physically inaccessible to an unauthorized

J.A. Onieva et al. (Eds.): WISTP 2008, LNCS 5019, pp. 136–149, 2008.

user. However in 1996 timing attacks were publicly introduced by Kocher in [Koc96]. Two years later he also introduced power analysis attacks with Jaffe and Jun [KJJ99]. Side Channel Analysis (SCA) is a group of techniques including simple power analysis (SPA) and differential power analysis (DPA). SCA threatens any naive cryptographic algorithm implementation. Since these first articles were published, power analysis has been widely investigated, some publications have focused on countermeasures and their drawbacks [FV03, MPO05, YLMH05] whereas others have focused on improving the efficiency of the attacks [ABDM00, BK03, BCO04].

One such countermeasure is the Binary with Random Initial Point (BRIP) algorithm(s) by Mamiya, Miyaji, Morimoto [MMM04] and Itoh, Izu and Takenada [IIT04], later improved in [IIT06]. BRIP countermeasure was originally designed for ECC and later extended to RSA cryptosystems. Its RSA variant corresponds to the countermeasure also proposed in [KHK+04] and is particularly interesting in terms of implementation as neither the bit size of the prime characteristic of the field is increased nor is the knowledge of the public exponent value needed.

Our study focuses on the exponentiation and for readability purposes, BRIP acronym will refer here to the BRIP RSA derivative of the countermeasure.

The paper is organized as follows: section 2 gives an overview of embedded asymmetric algorithms and their related side-channel potential vulnerabilities. Section 3 describes the BRIP algorithms with the current identified vulnerabilities and our implementation improvements. New attacks on these algorithms and recommendations will be presented in Section 4. We conclude our research in Section 5.

2   Power Analysis Background

Since the initial publication in [KJJ99] on Simple Power Analysis (SPA), many improvements have been made on this subject. Electronic devices, such as smart cards or other security products, are designed with thousands of logical gates switching differently depending on the executed operations and the data manipulated. The device power consumption of the chip depends on these operations which can be easily monitored and analysed on an oscilloscope. For instance, if the square operation has a different pattern on the power curve than the one for multiplication, it is obvious that the attacker can easily recover the secret exponent in a naive RSA implementation. Many other differences visible in the power curve can lead to the same kind of leakage on the private key(s). Developers must take into account all the potential vulnerabilities.

One of the first Collision Power Analysis attack is the Doubling Attack by Fouque and Valette [FV03]. It was applied on a scalar multiplication operation in ECC. They also explained how it could be extended to RSA implementations.

Differential Power Analysis (DPA) and its improvements represent the other main class of side channel attacks. The most well known is the Correlation Power Analysis (CPA) by Brier, Clavier and Olivier [BCO04]. It was later applied by Amiel, Feix and Villegas [AFV07] on most asymmetric algorithms. The first

DPA attack on RSA was done in 1999 by Messerges, Dabbish and Sloan [MDS99]. Enhanced DPA attacks, such as the Zero Value Point Attack published by Goubin in [Gou03], have also been done on elliptic curve implementations. Goubin's attack threatens Coron's randomization of the projective coordinates [Cor99] in the elliptic curve scalar multiplication. The first combination between fault injection and power analysis has also been applied to XTR in [CG04]. The BRIP algorithms have then been proposed to counteract the Zero Value Point Attack. Moreover, BRIP can also be applied in $GF(p)$ for cryptosystems based on the factorization and discrete logarithm problems, like RSA.

However Yen, Lien, Moon and Ha [YLMH05] presented a power collision attack on the BRIP countermeasure for RSA by using ± 1 values for input messages, and on the Square and Multiply Always algorithm by using $\pm m \mod n$ messages as an input for RSA.

3   Modular Exponentiations for BRIP Algorithms

Firstly we present the BRIP algorithm variant for RSA, we also introduce some improvements and optimizations for this countermeasure when combined with Montgomery modular multiplication.

3.1   Modular Multiplication and Exponentiation

We summarize the principles used later in this paper: modular multiplication and exponentiation, in particular the ones designed by Montgomery, which are particularly suitable for embedded implementations and the RSA public key cryptosystem.

3.2   Modular Multiplication

To compute modular multiplications $x \times y \mod n$ on long integers x, y and n Montgomery proposed the following efficient algorithm in [Mon85].

Montgomery Modular Multiplication
Given a modulus n and two integers x and y, of size v in base b, with $\gcd(n, b) = 1$ and $r = b^{\lceil log_b(n) \rceil}$, MontMul algorithm computes:

$$\mathsf{MontMul}(x,\, y,\, n) = x \times y \times r^{-1} \mod n$$

We suggest the reader to refer to Appendix A.1 and papers [Mon85] and [KAK96] if more detail on this operation is wished.

We can then use this operation to process efficiently Montgomery modular exponentiation (MontExp) as detailed in [Dhe98]. Compared to a classical Square and Multiply algorithm it consists of multiplying the message operand and the accumulator by $r \mod n$ before the exponentiation loop. In this case any intermediate result during the exponentiation is equal to $m^k.r \mod n$. At the end the r value is removed by doing a modular montgomery multiplication by 1. Refer to Appendix A.2 for the detailed algorithm.

Classical BRIP Implementation for RSA

Alg. 3.1 describes the classical BRIP implementation introduced in [KHK$^+$04] with a random v generated from a h-bit random seed u. It means $v = f(u)$. Value v must be as long as the modulus to prevent the implementation of a chosen message SPA. The security of the random value v is the same as the seed random u, this implies there are only 2^h possible values. For instance $v = (u|u \dots |u)$.

The major drawback is the time needed to compute the modulo inverse v^{-1} mod n. The next implementation avoids this if Montgomery modular multiplication hardware is available.

Algorithm 3.1. BRIP Exponentiation from left to right

INPUT: integers m and n such that $m < n$, k-bit exponent $d = (d_{k-1}d_{k-2} \dots d_1 d_0)_2$

OUTPUT: BRIP_Exp(m,d,n)$= m^d$ mod n

Step 1 If $m = 1$ **Return**(1)

Step 2 If $m = n - 1$ **Return**$((-1)^{d_0}$ mod $n)$

Step 3 Choose a random value v and compute v^{-1} mod n

Step 4 $a = v$, $m_0 = v^{-1}$ mod n, $m_1 = v^{-1}.m$ mod n

Step 5 for i from $k - 1$ to 0 **do**

$\qquad a = a \times a$ mod n

$\qquad a = a \times m_{d_i}$ mod n

Step 6 $a = a \times m_0$

Step 7 Return(a)

Second BRIP Implementation with MontMul

The inversion of random v mod n is a penalty for the BRIP algorithm performance. A solution consists in using the following property of the Montgomery multiplication: MontMul$(1, 1, n) = r^{-1}$ mod n. This gives an efficient way to compute an exponentiation with both a fixed base value (r) and a negative exponent. The idea is presented by Ciet and Feix in [CF05] and can also be applied to BRIP.

The v^{-1} mod n computation can be replaced by r^{-v} mod n implemented as an exponentiation with a relatively short exponent (typically $|v| << |d|$). This trick saves a lot of time compared to a modular inverse calculation.

Thus we obtained the Algorithm Alg. 3.2.

Step 5. of Alg. 3.2 replaces the costly inversion operation of random v in Alg. 3.1. However both previous algorithms Alg. 3.1 and Alg. 3.2 have a complexity of 2 which is the same as the well known Square and Multiply Always algorithm. Improvements can however be envisaged by using k-ary and sliding window methods [ÇKK]. In [MMM04] the authors also presented optimized versions of BRIP, one version is using the k-ary method.

Algorithm 3.2. MontExp-BRIP from left to right

INPUT: integers m and n such that $m < n$, k-bit exponent $d = (d_{k-1}d_{k-2}\ldots d_1 d_0)_2$
OUTPUT: MontExp$(m,d,n)= m^d \mod n$

Step 1 If $m = 1$ **Return**(1)

Step 2 If $m = n - 1$ **Return**$((-1)^{d_0} \mod n)$

Step 3 Choose a h-bit random value v

Step 4 Compute $V = r^v \mod n = $ MontExp(r,v,n)

Step 5 Compute $V_1 = $ MontMul$(1,1,n)$ and $U = r^{v^-} \mod n = $ MontExp(V_1,v,n)

Step 6 $a = V.r \mod n$, $m_1 = m.U.r \mod n$, $m_0 = U.r \mod n$

Step 7 for i from $k - 1$ to 0 **do**
   $a = $ MontMul(a, a, n)
   $a = $ MontMul(a, m_{d_i}, n)

Step 8 $a = $ MontMul(a, m_0, n)

Step 9 $a = $ MontMul$(a, 1, n)$

Step 10 Return(a)

Algorithm 3.3. MontExp-WBRIP from left to right

INPUT: integers m and n such that $m < n$, k-bit exponent $d = (d_{k-1}d_{k-2}\ldots d_1 d_0)_2$
OUTPUT: MontExp$(m,d,n)= m^d \mod n$

Step 1 If $m = 1$ **Return**(1)

Step 2 If $m = n - 1$ **Return**$((-1)^{d_0} \mod n)$

Step 3 Choose a h-bit random value v

Step 4 Compute $V = r^v \mod n = $ MontExp(r,v,n)

Step 5 Compute $U = r^{-3v} \mod n$

Step 6 Compute $a = V.r \mod n$, $m_0 = U.r \mod n$, $m_1 = m.U.r \mod n$

Step 7 Compute $m_2 = m^2.U.r \mod n$, $m_3 = m^3.U.r \mod n$

Step 8 for i from $k - 1$ to 0 by 2 **do**
   $a = $ MontMul(a, a, n)
   $a = $ MontMul(a, a, n)
   $a = $ MontMul$(a, m_{(2.d_i+d_{i-1})}, n)$

Step 9 $a = $ MontMul(a, m_0, n)

Step 10 $a = $ MontMul$(a, 1, n)$

Step 11 Return(a)

Algorithm Alg. 3.3 we present here, corresponds to WBRIP for RSA with MontMul. It corresponds to a 2-ary exponentiation with the BRIP countermeasure and the improvement we proposed with the Montgomery multiplication. There is no costly inversion operation and the algorithm complexity is 1.5, but more memory space is required for the pre-computations storage compared to both the previous versions.

In this case the mask value for computation is no longer r^{-v} but r^{-3v} as we manipulate scalar bits by 2-bit windows.

Depending on the memory contraints, the size of the window can be modified. For a k-bit window the algorithm complexity becomes equal to $1 + 1/k$.

4   Power Analysis Attacks on BRIP Like Algorithms

We present here an improvement to the power collision attack on RSA implementations based on the previous BRIP implementations. Fouque and Valette first [FV03] introduced power collision attacks on some of the classical elliptic curve scalar multiplication algorithms, they also explained how to extend the technique to modular exponentiations. Later Yen *et al.* [YLMH05] introduced collision power attacks based on chosen message values $\pm 1 \mod n$ that allows the secret exponent value d to be recovered from a single curve. Developpers must avoid BRIP computation when the input message equals $n - 1$ and simply return value 1 or $n - 1$ depending on the parity of the secret exponent.

In their article, some other variants of the attacks are presented, especially on the Square and Multiply Always algorithm by using $\pm m \mod n$ messages as input, but none of them compromise a full implementation of BRIP.

4.1   Collision Power Analysis on BRIP and MontExp-BRIP

Modular multiplication on a chip requires relatively long processing time and relatively high power consumption compared with symmetric algorithms, where for example, processing can be carried out in a few clock cycles in hardware implementations of AES.

In figure 1 we analyse power traces of the MonMult operation executed on a tamper resistant device such as a smart card.

We choose two different random messages m_1 and m_2 and for each message we execute three multiplications MontMul(m_1, m_1, n) and MontMul(m_2, m_2, n). We then collect the three power curves $C_{1,1}$, $C_{1,2}$ and $C_{1,3}$ of the multiplication with m_1 and three curves $C_{2,1}$, $C_{2,2}$ and $C_{2,3}$ of the multiplication with m_2.

We notice, cf. figure 1, that on the selected chip, the multiplication is a very power consuming operation. This is due to the large number of gates which are switching together in the asymmetric coprocessor logic.

From this curves we observe that power collisions occur for similar data manipulated by the chip. $C_{1,1}$, $C_{1,2}$ and $C_{1,3}$ are similar and have exactly the same power traces, as do $C_{2,1}$, $C_{2,2}$ and $C_{2,3}$. It means that $C_{i,j}$ collides with $C_{k,l}$ when $i = k$ while $C_{i,j}$ is different from $C_{k,l}$ when $i \neq k$.

Due to the important number of clock cycles in a modular multiplication in the power curve, we can assume that different input data will have different power trace patterns. This means we can distinguish collisions with a high probability. The tests we made on the selected chip confirm our assumption.

We analyse if we can exploit eventual collisions on the classical BRIP algorithm 3.1 and the MontExp-BRIP 3.2 for an h-bit random value v. For both algorithms the analysis will be identical.

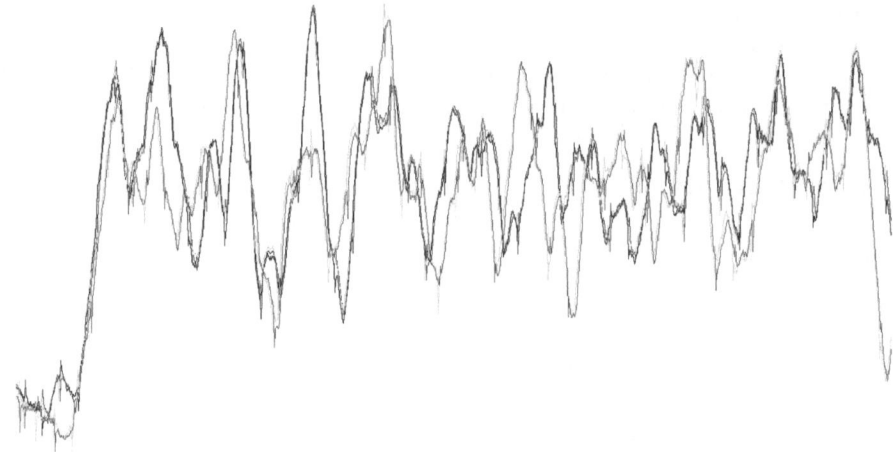

Fig. 1. Power consumption of a single modular multiplication for curves $C_{1,1}$, $C_{1,2}$, $C_{1,3}$ and $C_{2,1}$, $C_{2,2}$, $C_{2,3}$

Depending on the method for generating the random value v, it is obvious that in some cases, collisions on its values could happen when generating it. This will depend on the quality of the random and on its length. However for performance reasons BRIP and especially MontExp-BRIP can not use big values h.

We observe that if a colliding value for the random mask v appears, then by choosing as algorithm input message m for the first execution and $-m \mod n$ for the second one, we can have multiplications with similar operands in both executions. This could lead to distinguishable power collisions between the power curves of both executions. We then try to exploit these collisions to recover the secret exponent d.

Firstly we execute BRIP a number of times with input message m. For any execution a new random value v_1 is generated by the chip. Then we repeat this operation with the input message $-m \mod n$, for each execution a new random value v_2 is generated.

Let $d = d''.2^{i+1} + d_i.2^i + d'$ where, d_i is the current bit handled by the exponentiation loop, d'' the left part of d previously processed (left-to-right exponentiation) and d' the right remaining part of the exponent. In figure 2 we can observe for a step i of the BRIP execution what operands are manipulated by the chip for modular multiplications. In the first table we see these operand values during a real multiplication ($d_i = 1$), and in the second table when $d_i = 0$.

We detect collisions on any Fake Multiplication (multiplication by r^{-v}) operation when a collision happens on v and $v_1 = v_2$. Thus collision detection through power analysis is a real threat.

We can then observe on power traces when a collision occurs. We store in memory the power curves C_i of the BRIP execution with message m and C'_i with message $-m \mod n$. Then we search for two curves C_i and C'_j where

Message	Square	Message Multiplication ($d_i = 1$)	Square
m	$\left[m^{d''}.v_1\right]^2$	$\left[(m^{2.d''}).v_1^2\right] \times \left[m.v_1^{-1}\right]$	$\left[(m^{2.d''+1}).v_1\right]^2$
$-m$	$\left[(-m)^{d''}.v_2\right]^2$	$\left[(m^{2.d''}).v_2^2\right] \times \left[-m.v_2^{-1}\right]$	$\left[((-m)^{2.d''+1}).v_2\right]^2$
Collision if $v_1 = v_2$	-	No	No
Message	Square	Fake Multiplication ($d_i = 0$)	Square
m	$\left[m^{d''}.v_1\right]^2$	$\left[(m^{2.d''}).v_1^2\right] \times \left[v_1^{-1}\right]$	$\left[(m^{2.d''}).v_1\right]^2$
$-m$	$\left[(-m)^{d''}.v_2\right]^2$	$\left[(m^{2.d''}).v_2^2\right] \times \left[v_2^{-1}\right]$	$\left[(m^{2.d''}).v_2\right]^2$
Collision if $v_1 = v_2$	-	Yes	Yes

Fig. 2. BRIP execution for $d_i = 1$ and $d_i = 0$

Algorithm 4.4. BRIP Collision Attack

INPUT: $s = $ RSA-BRIP(m, d), $s' = $ RSA-BRIP$(-m, d)$
OUTPUT: Secret exponent d

Step 1 Choose a random value m in $[2, n-2]$.
Step 2 Collect k traces $(C_0, ..., C_{k-1})$ of BRIP execution with m as input message.
Step 3 Collect k traces $(C'_0, ..., C'_{k-1})$ of BRIP execution with $-m$ as input message.
Step 4 Find traces C_i and C'_j such as both traces are colliding on each BRIP Fake Multiply.
Step 5 Compute $S = |C_i - C'_j|$.
Step 6 Each non zero difference on S identify a true multiplication, i.e. $d_i = 1$

---

power collisions appear between the two curves. Then by subtracting C'_j to C_i we can recover the secret exponent d.

The probability of finding at least one colliding couple from both sets of k traces is approximated in [MOV96] (Fact 2.27) by:

$$p_{collision} \simeq 1 - e^{-((k^2)/|h|)}$$

where $|h|$ denotes the number of possible value for v so 2^h.

Figure 3 gives the probability of collision for a 32-bit random v relative to the number of encryptions done.

Thus in practice with 2^{32} possible values for v (32-bit random), two sets of $k = 78000$ curves are sufficient to have a probability of $\frac{1}{2}$ for obtaining a collision, where two sets of $k = 200000$ curves will lead to a collision and then a successful attack in 99 percents of cases.($p_{collision} \simeq 0.99$).

This collision attack is a serious threat and also appears on MontExp-BRIP, cf. Alg. 3.2. The random value v is not used in the same way in the algorithm but the analysis and the results of collision are similar.

We suggest using at least 96-bit random value v ($h = 96$ cf. figure 5) to prevent MontExp-BRIP and BRIP against such collision attacks. However it is obvious

h	k	collision
32	78000	0.507
32	$2^{17} \approx 131072$	0.864
32	161000	0.951
32	200000	0.990
32	$2^{18} \approx 262144$	0.999

Fig. 3. Probability of collision for $h = 32$

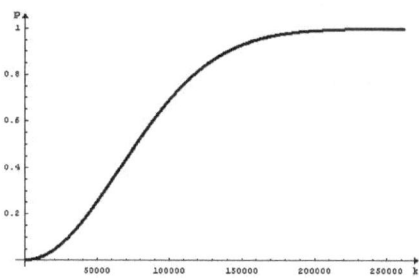

Fig. 4. Graph of probability of collision for a 32-bit value v

h	k	collision probability
16	$2^9 = 512$	0.864
16	$2^{10} = 1024$	0.999
64	5.1×10^9	0.505
64	2^{33}	0.864
64	2^{34}	0.999
96	3.3×10^{14}	0.497
96	2^{48}	0.864
96	2^{49}	0.999

Fig. 5. Probability of collision for other h values

that such random lengths will reduce the performance of these implementations and then the reason for using them.

4.2   Collision Attack on MontExp-WBRIP

We analyse the impact of power collisions in the MontExp-WBRIP implementation. In the figure 6 we replace the notation MontMul by MM.

Let $d = d''.2^{i+1} + d_i.2^i + d_{i-1}.2^{i-1} + d'$ where d_i and d_{i-1} are the two bits of the 2-bit window handled for each exponentiation loop, d'' the left part of d previously processed and d' the right part being process in next steps.

Message	Square	Square	$M_{(2d_i+d_{i-1}=0)}$ MontMul
m	$MM(m^{d''}.r^{v_1}, m^{d''}.r^{v_1})$	$MM(m^{2.d''}.r^{v_1}, m^{2.d''}.r^{v_1})$	$MM(m^{4.d''}.r^{4v_1}, r^{-3v_1})$
$-m$	$MM((-m)^{d''}.r^{v_2}, (-m)^{d''}.r^{v_2})$	$MM(m^{2.d''}.r^{v_2}, m^{2.d''}.r^{v_2})$	$MM(m^{4.d''}.r^{4v_2}, r^{-3v_2})$
If $v_1=v_2$	-	Yes	Yes
Message	Square	Square	$M_{(2d_i+d_{i-1}=1)}$ MontMul
m	$MM(m^{d''}.r^{v_1}, m^{d''}.r^{v_1})$	$MM(m^{2.d''}.r^{v_1}, m^{2.d''}.r^{v_1})$	$MM(m^{4.d''}.r^{4v_1}, m.r^{-3v_1})$
$-m$	$MM((-m)^{d''}.r^{v_2}, (-m)^{d''}.r^{v_2})$	$MM(m^{2.d''}.r^{v_2}, m^{2.d''}.r^{v_2})$	$MM(m^{4.d''}.r^{4v_2}, (-m).r^{-3v_2})$
If $v_1=v_2$	-	Yes	No
Message	Square	Square	$M_{(2d_i+d_{i-1}=2)}$ MontMul
m	$MM(m^{d''}.r^{v_1}, m^{d''}.r^{v_1})$	$MM(m^{2.d''}.r^{v_1}, m^{2.d''}.r^{v_1})$	$MM(m^{4.d''}.r^{4v_1}, m^2.r^{-3v_1})$
$-m$	$MM((-m)^{d''}.r^{v_2}, (-m)^{d''}.r^{v_2})$	$MM(m^{2.d''}.r^{v_2}, m^{2.d''}.r^{v_2})$	$MM(m^{4.d''}.r^{4v_2}, m^2.r^{-3v_2})$
If $v_1=v_2$	-	Yes	Yes
Message	Square	Square	$M_{(2d_i+d_{i-1}=3)}$ MontMul
m	$MM(m^{d''}.r^{v_1}, m^{d''}.r^{v_1})$	$MM(m^{2.d''}.r^{v_1}, m^{2.d''}.r^{v_1})$	$MM(m^{4.d''}.r^{4v_1}, m^3.r^{-3v_1})$
$-m$	$MM((-m)^{d''}.r^{v_2}, (-m)^{d''}.r^{v_2})$	$MM(m^{2.d''}.r^{v_2}, m^{2.d''}.r^{v_2})$	$MM(m^{4.d''}.r^{4v_2}, (-m)^3.r^{-3v_2})$
If $v_1=v_2$	-	Yes	No

Fig. 6. WBRIP execution for possible $2d_i + d_{i-1}$ values

We can observe in figure 6 the different possible collisions on curves when random values v and v_1 collide. But in WBRIP it does not give us as much information as in the previous algorithms. The collisions will indicate that the 2-bit window value is either 00 or 10 so $d_{i-1} = 0$ and non collisions will indicate the 2-bit window value is either 01 or 11 so $d_{i-1} = 1$. Then we recover here half of the bits of the secret exponent d.

Indeed, we can extend this result to any k-ary implementation of BRIP exponentiation as the Collision Attack gives the information on the parity of i in $m^i.r^{-2^{k-1}.v}$ operand used during the Multiplication operation. Therefore the number of bits recovered by a collision is equal to : $|d|/k$, namely $|d|$ for $k = 1$, $|d|/2$ for $k = 2$ and so on.

Thus we also suggest using at least 96-bit random value v to prevent MontExp-WBRIP from such collision attacks.

4.3   Collision Attack of BRIP Implementations for RSA CRT

These collision attacks can be similarly applied to RSA CRT exponentiations protected with MontExp-BRIP, MontExp-WBRIP or BRIP algorithms.

Indeed when $n = p.q$, p and q being prime numbers of equivalent lengths, choosing $\pm m$ mod n messages leads to the manipulation of $\pm m$ mod p and $\pm m$ mod q in the CRT exponentiations once the reductions by p and q have been done.

Then the previous collision analysis applies identically to RSA CRT using any of the previous BRIP algorithms.

4.4 Implementing MontExp-BRIP Countermeasure

We notice that both exponentiations: $r^v \mod n = \mathsf{MontExp}(r, v, n)$ and $r^{-v} \mod n = \mathsf{MontExp}(V_1, v, n)$ need to be carefully implemented against the classical power analysis techniques. Indeed it is obvious that if v is recovered, each operand value in algorithm 3.2 becomes deterministic and then statistical attacks can be envisaged to recover the secret exponent.

The most important threat is Timing Attack (TA) for which Double and Add Always or Side Channel Atomicity [CCJ04] are both convenient countermeasures.

Anyway, protection against TA may not be sufficient as the operation $r^{-v} = \mathsf{MontExp}(V_1, v, n)$ can be sensitive to SPA. This is due to the particular Hamming Weight of one of the Multiply operands, explicitly $\overline{m}$ in $a = \mathsf{MonMul}(a, \overline{m}, n)$ with $\overline{m} = f_n(m) = r^{-1} * r = 1$.

Analysing the implementation details of MontMul gives some clues to explaining the leakage. During the computation of $\mathsf{MontMul}(a, 1, n)$, we notice than most of multiplications involved in Step 2 of algorithm A.5 are composed of integer multiplications by 0 or 1 which have a straightforward impact on the power consumption by significantly lowering it compared to the multiplication of two random operands. It can then be feasible to deduce directly from the power curve the nature of each operation and recover v value for each curve.

A simple tweak to counteract such an SPA attack is to compute $(-r)^{-v}$ rather than r^{-v}, $n-1$ will then replace 1 as input operand of MontMult during Multiply operations. This may still not be sufficient to protect against advanced SPA or Template Analysis attacks as intrinsically r^{-v} or r^v exponentiations are not randomized.

Applying additional randomization techniques on r^{-v} and r^v exponentiations could be envisaged to protect against such threats but will reduce efficiency and at the same time reason for MontExp-BRIP countermeasures.

5 Conclusion

Several possible implementations of BRIP algorithms have been presented in this paper. We used the efficiency of Montgomery modular arithmetic to provide an efficient message masking technique. We showed with these implementations detecting collisions through power analysis, and especially during modular multiplications, is a realistic threat. We also explained that random length must be chosen very carefully to prevent these implementations from the collision attacks we have described. Thus using 32-bit or even 64-bit random values should be avoided here. In the case where ISO random padding is used, it naturally prevents our implementation from this collision attack and allows a shorter random value (32 bits) to be used, but it is not always the case.

We also stress to the reader that random value manipulation must be strongly protected in the MontExp-BRIP algorithm against the different side channel techniques in order to prevent the random recovery by power leakage. Such random recovery could then lead to other classical power analysis on the secret.

Acknowledgements

The authors would like to thank Mathieu Ciet for its comments, and Sean Commercial for his help for the final version.

References

[ABDM00]   Akkar, M.-L., Bevan, R., Dischamp, P., Moyart, D.: Power Analysis, What Is Now Possible. In: Okamoto, T. (ed.) ASIACRYPT 2000. LNCS, vol. 1976, pp. 489–502. Springer, Heidelberg (2000)
[AFV07]   Amiel, F., Feix, B., Villegas, K.: Power Analysis for Secret Recovering and Reverse Engineering of Public Key Algorithms. In: Selected Areas in Cryptography. LNCS, vol. 4876, pp. 110–125. Springer, Heidelberg (2007)
[BCO04]   Brier, E., Clavier, C., Olivier, F.: Correlation Power Analysis with a Leakage Model. In: Joye, M., Quisquater, J.-J. (eds.) CHES 2004. LNCS, vol. 3156, pp. 16–29. Springer, Heidelberg (2004)
[BK03]   Bevan, R., Knudsen, E.: Ways to Enhance Differential Power Analysis. In: Lee, P.J., Lim, C.H. (eds.) ICISC 2002. LNCS, vol. 2587, pp. 327–342. Springer, Heidelberg (2003)
[CCJ04]   Chevallier-Mames, B., Ciet, M., Joye, M.: Low-cost solutions for preventing simple side-channel analysis: side-channel atomicity. IEEE Transactions on Computers 53(6), 760–768 (2004)
[CF05]   Ciet, M., Feix, B.: Cryptographic method comprising a modular exponentiation secured against hidden-channel attacks, crypto processor for implementing the method and associated chip card. Gemplus Patent WO2007074149 (2005)
[CG04]   Ciet, M., Giraud, C.: Transient fault induction attacks on XTR. In: López, J., Qing, S., Okamoto, E. (eds.) ICICS 2004. LNCS, vol. 3269, pp. 440–451. Springer, Heidelberg (2004)
[ÇKK]   Koç, K.: Analysis of Sliding Window Techniques for Exponentiation. Computers and Mathematics with Applications 30(10), 17–24 (1995)
[Cor99]   Coron, J.-S.: Resistance against differential power analysis for elliptic curve cryptosystems. In: Koç, Ç.K., Paar, C. (eds.) CHES 1999. LNCS, vol. 1717, pp. 292–302. Springer, Heidelberg (1999)
[DH76]   Diffie, W., Hellman, M.E.: New Directions in cryptography. IEEE Transactions on Information Theory 22(6), 644–654 (1976)
[Dhe98]   Dhem, J.-F.: Design of an efficient public-key cryptographic library for RISC-based smart cards. PhD thesis, Université catholique de Louvain, Louvain (1998)
[FV03]   Fouque, P.-A., Valette, F.: The Doubling Attack - why upwards is better than downwards. In: Walter, C.D., Koç, Ç.K., Paar, C. (eds.) CHES 2003. LNCS, vol. 2779, pp. 269–280. Springer, Heidelberg (2003)
[Gou03]   Goubin, L.: A refined power-analysis attack on elliptic curve cryptosystems. In: Desmedt, Y.G. (ed.) PKC 2003. LNCS, vol. 2567, pp. 199–210. Springer, Heidelberg (2002)

[IIT04]   Itoh, K., Izu, T., Takenaka, M.: Efficient Countermeasures against Power Analysis for Elliptic Curve Cryptosystems. In: Quisquater, J.-J., Paradinas, P., Deswarte, Y., El Kalam, A.A. (eds.) CARDIS, pp. 99–114. Kluwer, Dordrecht (2004)

[IIT06]   Itoh, K., Izu, T., Takenaka, M.: Improving the Randomized Initial Point Countermeasure against DPA. In: Zhou, J., Yung, M., Bao, F. (eds.) ACNS 2006. LNCS, vol. 3989, pp. 459–469. Springer, Heidelberg (2006)

[KAK96]   Koç, Ç.K., Acar, T., Kaliski, B.-S.: Analysing and comparing Montgomery multiplication algorithms. IEEE Micro. 16(3), 26–33 (1996)

[KHK⁺04] Kim, C., Ha, J., Kim, S.-H., Kim, S., Yen, S.-M., Moon, S.-J.: A Secure and Practical CRT-Based RSA to Resist Side Channel Attacks. In: Laganá, A., Gavrilova, M.L., Kumar, V., Mun, Y., Tan, C.J.K., Gervasi, O. (eds.) ICCSA 2004. LNCS, vol. 3043, pp. 150–158. Springer, Heidelberg (2004)

[KJJ99]   Kocher, P.C., Jaffe, J., Jun, B.: Differential Power Analysis. In: Wiener, M.J. (ed.) CRYPTO 1999. LNCS, vol. 1666, pp. 388–397. Springer, Heidelberg (1999)

[Kob87]   Koblitz, N.: Elliptic curve cryptosystems. Math. of Comp. 48(177), 203–209 (1987)

[Koc96]   Kocher, P.C.: Timing attacks on implementations of Diffie-Hellman, RSA, DSS, and other systems. In: Koblitz, N. (ed.) CRYPTO 1996. LNCS, vol. 1109, pp. 104–113. Springer, Heidelberg (1996)

[MDS99]   Messerges, T.S., Dabbish, E.A., Sloan, R.H.: Power analysis attacks of modular exponentiation in smartcards. In: Koç, Ç.K., Paar, C. (eds.) CHES 1999. LNCS, vol. 1717, pp. 144–157. Springer, Heidelberg (1999)

[Mil86]   Miller, V.S.: Use of elliptic curves in cryptography. In: Williams, H.C. (ed.) CRYPTO 1985. LNCS, vol. 218, pp. 489–502. Springer, Heidelberg (1986)

[MMM04]   Mamiya, H., Miyaji, A., Morimoto, H.: Efficient Countermeasures against RPA, DPA, and SPA. In: Joye, M., Quisquater, J.-J. (eds.) CHES 2004. LNCS, vol. 3156, pp. 343–356. Springer, Heidelberg (2004)

[Mon85]   Montgomery, P.L.: Modular multiplication without trial division. Mathematics of Computation 44(170), 519–521 (1985)

[MOV96]   Menezes, A., van Oorschot, P.C., Vanstone, S.A.: Handbook of Applied Cryptography. CRC Press, Boca Raton (1996)

[MPO05]   Mangard, S., Pramstaller, N., Oswald, E.: Successfully attacking masked AES hardware implementations. In: Rao, J.R., Sunar, B. (eds.) CHES 2005. LNCS, vol. 3659, pp. 157–171. Springer, Heidelberg (2005)

[RSA78]   Rivest, R.L., Shamir, A., Adleman, L.: A method for obtaining digital signatures and public-key cryptosystems. Communications of the ACM 21, 120–126 (1978)

[YLMH05] Yen, S.-M., Lien, W.-C., Moon, S., Ha, J.: Power Analysis by Exploiting Chosen Message and Internal Collisions - Vulnerability of Checking Mechanism for RSA-decryption. In: Dawson, E., Vaudenay, S. (eds.) Mycrypt 2005. LNCS, vol. 3715, pp. 183–1956. Springer, Heidelberg (2005)

A   Montgomery Arithmetic

A.1   Montgomery Multiplication

Algorithm A.5. MontMul: Montgomery modular multiplication algorithm

INPUT: n, $0 \leq x = (x_{v-1}x_{v-2}\ldots x_1x_0)_b, y = (y_{v-1}y_{v-2}\ldots y_1y_0)_b \leq n - 1$, $n' = -n^{-1} \mod b$

OUTPUT: $x \times y \times r^{-1} \mod n$

Step 1 $a = (a_{v-1}a_{v-2}\ldots a_1a_0) \leftarrow 0$

Step 2 for i from 0 to $v - 1$ **do**

$u_i \leftarrow (a_0 + x_i \times y_0) \times n' \mod b$

$a \leftarrow (a + x_i \times y + u_i \times n)/b$

Step 3 if $a \geq n$ **then** $a \leftarrow a - n$

Step 4 Return(a)

A.2   Montgomery Exponentiation

Algorithm A.6. MontExp: Montgomery Square and Multiply from left to right

INPUT: integers m and n such that $m < n$, k-bit exponent $d = (d_{k-1}d_{k-2}\ldots d_1d_0)_2$

OUTPUT: MontExp(m,d,n)$= m^d \mod n$

Step 1 $a = r$

Step 2 $\overline{m} = m \times r \mod n$

Step 3 for i from $k - 1$ to 0 **do**

$a = \mathsf{MontMul}(a,a,n)$

if $d_i = 1$ **then** $a = \mathsf{MontMul}(a,\overline{m},n)$

Step 4 $a = a \times r^{-1} \mod n = \mathsf{MontMul}(a,1,n)$

Step 5 Return(a)

Author Index

Lecture Notes in Computer Science

Sublibrary 4: Security and Cryptology

For information about Vols. 1– 3866
please contact your bookseller or Springer

Vol. 4462: D. Sauveron, K. Markantonakis, A. Bilas, J.-J. Quisquater (Eds.), Information Security Theory and Practices. XII, 255 pages. 2007.

Vol. 4450: T. Okamoto, X. Wang (Eds.), Public Key Cryptography – PKC 2007. XIII, 491 pages. 2007.

Vol. 4437: J.L. Camenisch, C.S. Collberg, N.F. Johnson, P. Sallee (Eds.), Information Hiding. VIII, 389 pages. 2007.

Vol. 4392: S.P. Vadhan (Ed.), Theory of Cryptography. XI, 595 pages. 2007.

Vol. 4377: M. Abe (Ed.), Topics in Cryptology – CT-RSA 2007. XI, 403 pages. 2006.

Vol. 4356: E. Biham, A.M. Youssef (Eds.), Selected Areas in Cryptography. XI, 395 pages. 2007.

Vol. 4341: P.Q. Nguyên (Ed.), Progress in Cryptology - VIETCRYPT 2006. XI, 385 pages. 2006.

Vol. 4332: A. Bagchi, V. Atluri (Eds.), Information Systems Security. XV, 382 pages. 2006.

Vol. 4329: R. Barua, T. Lange (Eds.), Progress in Cryptology - INDOCRYPT 2006. X, 454 pages. 2006.

Vol. 4318: H. Lipmaa, M. Yung, D. Lin (Eds.), Information Security and Cryptology. XI, 305 pages. 2006.

Vol. 4307: P. Ning, S. Qing, N. Li (Eds.), Information and Communications Security. XIV, 558 pages. 2006.

Vol. 4301: D. Pointcheval, Y. Mu, K. Chen (Eds.), Cryptology and Network Security. XIII, 381 pages. 2006.

Vol. 4300: Y.Q. Shi (Ed.), Transactions on Data Hiding and Multimedia Security I. IX, 139 pages. 2006.

Vol. 4298: J.K. Lee, O. Yi, M. Yung (Eds.), Information Security Applications. XIV, 406 pages. 2007.

Vol. 4296: M.S. Rhee, B. Lee (Eds.), Information Security and Cryptology – ICISC 2006. XIII, 358 pages. 2006.

Vol. 4284: X. Lai, K. Chen (Eds.), Advances in Cryptology – ASIACRYPT 2006. XIV, 468 pages. 2006.

Vol. 4283: Y.Q. Shi, B. Jeon (Eds.), Digital Watermarking. XII, 474 pages. 2006.

Vol. 4266: H. Yoshiura, K. Sakurai, K. Rannenberg, Y. Murayama, S.-i. Kawamura (Eds.), Advances in Information and Computer Security. XIII, 438 pages. 2006.

Vol. 4258: G. Danezis, P. Golle (Eds.), Privacy Enhancing Technologies. VIII, 431 pages. 2006.

Vol. 4249: L. Goubin, M. Matsui (Eds.), Cryptographic Hardware and Embedded Systems - CHES 2006. XII, 462 pages. 2006.

Vol. 4237: H. Leitold, E.P. Markatos (Eds.), Communications and Multimedia Security. XII, 253 pages. 2006.

Vol. 4236: L. Breveglieri, I. Koren, D. Naccache, J.-P. Seifert (Eds.), Fault Diagnosis and Tolerance in Cryptography. XIII, 253 pages. 2006.

Vol. 4219: D. Zamboni, C. Krügel (Eds.), Recent Advances in Intrusion Detection. XII, 331 pages. 2006.

Vol. 4189: D. Gollmann, J. Meier, A. Sabelfeld (Eds.), Computer Security – ESORICS 2006. XI, 548 pages. 2006.

Vol. 4176: S.K. Katsikas, J. López, M. Backes, S. Gritzalis, B. Preneel (Eds.), Information Security. XIV, 548 pages. 2006.

Vol. 4117: C. Dwork (Ed.), Advances in Cryptology - CRYPTO 2006. XIII, 621 pages. 2006.

Vol. 4116: R. De Prisco, M. Yung (Eds.), Security and Cryptography for Networks. XI, 366 pages. 2006.

Vol. 4107: G. Di Crescenzo, A. Rubin (Eds.), Financial Cryptography and Data Security. XI, 327 pages. 2006.

Vol. 4083: S. Fischer-Hübner, S. Furnell, C. Lambrinoudakis (Eds.), Trust and Privacy in Digital Business. XIII, 243 pages. 2006.

Vol. 4064: R. Büschkes, P. Laskov (Eds.), Detection of Intrusions and Malware & Vulnerability Assessment. X, 195 pages. 2006.

Vol. 4058: L.M. Batten, R. Safavi-Naini (Eds.), Information Security and Privacy. XII, 446 pages. 2006.

Vol. 4047: M.J.B. Robshaw (Ed.), Fast Software Encryption. XI, 434 pages. 2006.

Vol. 4043: A.S. Atzeni, A. Lioy (Eds.), Public Key Infrastructure. XI, 261 pages. 2006.

Vol. 4004: S. Vaudenay (Ed.), Advances in Cryptology - EUROCRYPT 2006. XIV, 613 pages. 2006.

Vol. 3995: G. Müller (Ed.), Emerging Trends in Information and Communication Security. XX, 524 pages. 2006.

Vol. 3989: J. Zhou, M. Yung, F. Bao (Eds.), Applied Cryptography and Network Security. XIV, 488 pages. 2006.

Vol. 3969: Ø. Ytrehus (Ed.), Coding and Cryptography. XI, 443 pages. 2006.

Vol. 3958: M. Yung, Y. Dodis, A. Kiayias, T. Malkin (Eds.), Public Key Cryptography - PKC 2006. XIV, 543 pages. 2006.

Vol. 3957: B. Christianson, B. Crispo, J.A. Malcolm, M. Roe (Eds.), Security Protocols. IX, 325 pages. 2006.

Vol. 3956: G. Barthe, B. Grégoire, M. Huisman, J.-L. Lanet (Eds.), Construction and Analysis of Safe, Secure, and Interoperable Smart Devices. IX, 175 pages. 2006.

Vol. 3935: D.H. Won, S. Kim (Eds.), Information Security and Cryptology - ICISC 2005. XIV, 458 pages. 2006.

Vol. 3934: J.A. Clark, R.F. Paige, F.A.C. Polack, P.J. Brooke (Eds.), Security in Pervasive Computing. X, 243 pages. 2006.

Vol. 3928: J. Domingo-Ferrer, J. Posegga, D. Schreckling (Eds.), Smart Card Research and Advanced Applications. XI, 359 pages. 2006.

Vol. 3919: R. Safavi-Naini, M. Yung (Eds.), Digital Rights Management. XI, 357 pages. 2006.

Vol. 3903: K. Chen, R. Deng, X. Lai, J. Zhou (Eds.), Information Security Practice and Experience. XIV, 392 pages. 2006.

Vol. 3897: B. Preneel, S. Tavares (Eds.), Selected Areas in Cryptography. XI, 371 pages. 2006.

Vol. 3876: S. Halevi, T. Rabin (Eds.), Theory of Cryptography. XI, 617 pages. 2006.